D1286809

HYPERSEA

M. C. Escher, 1942, Study of Plants, India ink.
© 1942 M. C. ESCHER/CORDON ART, BAARN, HOLLAND.

HYPERSEA

Life on Land

Mark A. S. McMenamin and
Dianna L. S. McMenamin

Columbia University Press
New York

QH360.5
.M36
1994

Columbia University Press
New York   Chichester, West Sussex
All rights reserved

Library of Congress Cataloging-in-Publication Data
McMenamin, Mark A.
    Hypersea : life on land / Mark A. S. McMenamin and Dianna L. S.
McMenamin.
        p.   cm.
    Includes bibliographical references and index.
    ISBN 0–231–07530–8
    1. Hypersea.   2. Symbiogenesis.   I. McMenamin, Dianna L.
Schulte.
    II. Title.
    QH360.5.M36   1994
    575—dc20   94–15324
                                                        CIP

Casebound editions of Columbia University Press books are
printed on permanent and durable acid-free paper.

c 10 9 8 7 6 5 4 3 2 1

To Sarah and Amy

CONTENTS

I don't know what I may seem to the world, but as to myself, I seem to have been only like a boy playing on the seashore and diverting myself in now and then finding a smoother pebble or a prettier shell than ordinary, whilst the great ocean of truth lay all undiscovered before me.

—Sir Isaac Newton

FOREWORD

Lynn Margulis

Hypersea as a "physical entity" is the connectedness of life on land: from fungal-entangled roots to treetops, flitting moths, springing locusts, and soaring hawks. In this Vernadskian view, living matter has extended the marine over the land bringing phosphorus, water, and horizontal throbbing to the barren dryness of the continents. Without Hypersea these raised portions of inexorably moving plates would languish in the scorching sun as, changing latitude and longitude, their granitic peaks surged up only to be later weathered down.

Hypersea, the McMenamins argue, is also scientific theory. A specific application of new scientific observations, it blends Vernadskian biospheric thinking and Lovelockian Gaian spatial connectedness with the long temporal view of Darwin. All life on Earth shows common ancestry. The plants no more "conquered the land" alone than Christopher Columbus "discovered America" alone. With Ferdinand and Isabella behind the scenes and a land in which thousands of Nahuat and Mayan speakers, some already urbanized, enjoyed a culture that would strongly influence our own, the intrepid Genovese sailor is a convenient mythical symbol for the many other forms of life that accompanied him. The Spanish ships returned to the Europe that sustained them with *Zea mays*, *Nicotiana*, tomato, and turkey. So too the lignified rhyniophytes and subsequent seed ferns and cycads are mostly symbols.

Far more occurred than just the sweep of forest trees as plants, over sterile rock and soil, "invaded the land." Indeed, as the McMenamins explain here, it was the expansion of "animated water," Hypersea, as a single entity that colonized the portions of the plates that had emerged from the sea. The celebration, long overdue, of a fifth kingdom, the fungi, is played out here as well. The McMenamin geological perspective, displaying that fungal undergrowth upon which we all depend, deserves close consideration, especially by biological scientists. Without fungi the plants could never have come ashore. In the days when the "information superhighway" is the buzz phrase, we would do well to look at our inventive fungal predecessors who, for four hundred million years, have already been leading the communication network of life on land.

In short, we have been preempted and preceded by that same Hypersea of which we represent a recent and rapidly expanding part. These Hypersea links to ancient ocean and regolith are worthy of the attention that the McMenamins advocate. We now look wide-eyed upon a land surface whose history we thought we understood. Returning to where we stood before, enlightened by a century of biological and paleontological insight, we now see this place for the first time.

PREFACE

We like to think of ourselves as concept people. In accordance with our early training, Mark's concepts tend to be geological and Dianna's tend to be biological. We both love paleontology because it is the marriage of the biological and the geological.

When we wandered into the Brattleboro Public Library in July of 1988, Mark noticed an article about Vladimir Vernadsky (1863–1945), Russian geologist and geochemist. One particular Vernadsky quotation hit us both like a bolt out of the blue:

> There will always be scientists who clearly feel and embrace our planet's living nature, which is full of the perpetual beat of life. For them, understanding the oneness of nature is the guiding thread of all of their scientific research.

Vernadsky had articulated an aspect of the earth sciences we find exciting. Some months later we compared notes with Lynn Margulis, who had recently moved to the University of Massachusetts at Amherst. Margulis had a deep appreciation for the importance of Vernadsky's work and had begun highlighting it in her lectures and writings.

Vernadsky's research program has been greatly amplified both by Margulis and by the research and writing of the distinguished scientist James Lovelock, whose work has helped to bring the perpetual beat of planetary life to the attention of millions of people. Thanks to his efforts, scientists and science watchers have begun to realize the impact of life as a geological and planetary force.

The research of these great bio-geo-chemists focuses on the relationships between things biological and things geological. Biological manipulation of matter, symbiotic connections between organisms, and interlinked feedback processes infuse the writings of Vernadsky, Margulis, and Lovelock.

In this book we go beyond consideration of the connections, even the intimate ones, between geology and biology. For us a major part of the earth's surface is *literally* alive. This living surface feature is widespread, it is massive, and it includes our own bodies. It has not been given its proper due before because no one has seen, so to speak, the forest for the trees. We call this living landform Hypersea, and we invite you to join with us in its exploration.

Mark McMenamin
Dianna McMenamin

ACKNOWLEDGMENTS

We thank D. M. Ball, F. Collier, S. Conway Morris, P. R. Crane, D. Edwards, A. M. Ellison, M. S. Fuller, P. Gill, J. Gray, F. Grober, S. E. and P. Gruber, F. Howarth, M. Johnson, J. Kingeter, C. LaBandiera, A. Lapo, P. Mankiewicz, S. E. Miller, D. O'Shea, G. Osmer, S. Ostroumov, V. Page, S. Rachootin, F. D. Reed, S. M. Rowland, B. Rubinstein, D. Schwartzman, A. Seilacher, N. Sheats, M. Sogin, P. K. Strother, T. N. Taylor, and H. Winkelmann for assistance with numerous aspects of the research, preparation, and writing of this book.

Lynn Margulis deserves special mention for her help with identifying organisms and for discussion of the marvels of the Protoctista. Edward Lugenbeel offered his unfaltering support as editor at crucial moments. Connie Barlow provided encouragement and expert editorial assistance on the manuscript. Ian K. Ross first opened our eyes to the strange and wonderful nature of Kingdom Fungi.

Research has been supported in part by a Presidential Young Investigator Grant #EAR 8857995 from the National Science Foundation.

HYPERSEA

ONE

Introduction to Hypersea

The geologic time scale depicted in figure 1.1 is familiar to anyone interested in earth history. It is, however, misleading because it represents only about 12 percent of the entire 4.5-billion-year span of geologic time. The comings and goings of trilobites, mollusks, and other relatively complex organisms did make it possible for early stratigraphers to recognize change in the fossil record and to correlate in time outcrops separated by hundreds or thousands of miles. Fossils of complex organisms, moreover, were crucially important in the invention of the idea of geologic time as well as in its translation into specific periods, such as the Devonian and Jurassic. But a continuing fixation on such easily recognized and charismatic fauna has also hindered an understanding of evolution.

When the geologic scale is viewed in its entirety, the first appearance of animals and plants is nearly simultaneous. Indeed, even when viewed from the perspective of the time scale of life—which began with bacteria at least 3.5 billion years ago—the appearance of multicellular plants and animals seems eerily synchronous. Terrestrial plants as well as marine animals with shells burst into global abundance over a span of only a few tens of millions of years. Shelled animals appeared about 540 million years ago, at the beginning of the Paleozoic, and plants evolved about 480 million years ago. Is it pure coincidence that at least two wholly new kingdoms of life (out of a living pool that before them included only bacteria and algal or

FIGURE 1.1

The Geologic Time Scale. A. The eras, periods, and epochs of the last 600 million years. B. The periods, epochs, and ages of the Paleozoic era.

ERA	PERIOD	EPOCH	Millions of Years Ago
	Quaternary	Holocene	
		Pleistocene	1.8
CENOZOIC	Neogene	Pliocene	
		Miocene	24
	Paleogene	Oligocene	
		Eocene	
		Paleocene	66
MESOZOIC	Cretaceous	Late	
		Early	144
	Jurassic	Late	
		Middle	
		Early	208
	Triassic	Late	
		Middle	
		Early	245
PALEOZOIC	Permian	Late	
		Early	286
	Carboniferous (Pennsylvanian)	Late	320
	Carboniferous (Mississippian)	Early	360
	Devonian	Late	374
		Middle	387
		Early	408
	Silurian	Late	421
		Early	438
	Ordovician	Late	458
		Middle	478
		Early	505
	Cambrian	Late	
		Middle	
		Early	~540
PROTEROZOIC or "Late Precambrian"	Vendian		~580

The Paleozoic Era

PERIOD	EPOCH	AGE	Millions of Years Ago
Permian	Late		
	Early		286
Carboniferous (Pennsylvanian)	Late	Gzelian/Kasimovian	296
		Moscovian	312
		Bashkirian	320
Carboniferous (Mississippian)	Early	Serpukhovian	333
		Visean	352
		Tournaisian	360
Devonian	Late	Famennian	367
		Frasnian	374
	Middle	Givetian	380
		Eifelian	387
	Early	Emsian	394
		Seigenian	401
		Gedinnian	408
Silurian	Late	Pridolian	414
		Ludlovian	421
	Early	Wenlockian	428
		Llandoverian	438
Ordovician	Late	Ashgillian	448
		Caradocian	458
	Middle	Llandeilan	468
		Llanvirnian	478
	Early	Arenigian	488
		Tremadocian	505
Cambrian	Late		
	Middle		
	Early		~540

amoeba-like protoctists) arose almost simultaneously? Or was there a common cause?

A number of hypotheses have recently been invoked to explain these events. One idea is that the origin of animals and the origin of plants are both consequences of the first appearance of cells with the ability to arrange themselves into multicellular organisms. A second idea attributes animal and plant origins to rising levels of atmospheric and oceanic oxygen: the ambient concentrations of oxygen crossed a threshold that made it possible for oxygen-dependent cells to survive even when their membranes were buried under layers of sibling cells, rather than being directly exposed to the environment. A third hypothesis sees global ecological changes, such as the appearance of the first predators, as a trigger for the sudden evolution of plants and animals.

In this book we present an explanation for both the emergence of land animals and the origin of the plant kingdom (virtually all of which lives on land). Our explanation flows out of a new conception of the surface of Earth called Hypersea.

Why Hypersea?

Hypersea, as we define it, is made up of most but not all of the land biota. Specifically, *Hypersea is the sum total of plant, animal, protoctistan, and fungal life on land, plus any viral or bacterial symbionts or parasites living in intimate association with the tissues of these organisms.*[1] Hypersea does not include the bacterial component of the terrestrial biota that was able to live on land long before the first land plant appeared. Hypersea does not therefore include the kinds of bacterial communities, such as microbial mats and rock varnish microbes, that flourished before the invasion of their habitat by plants, animals, and fungi.

Why is this distinction important? Fundamentally, to be

part of Hypersea, a living organism must actively direct a flow of nutrient-rich fluids or be intimately associated with one that does. This is because, in our view, the most awesome aspect of the land biota is not that it includes both giant redwood trees and tiny parasitoid wasps, but that it vastly surpasses the marine biota in the physical connectedness of its constituents. The organisms of the land biota found that intimacy was a way to survive in an almost impossibly harsh environment. Lichens, which are found in landscapes as hostile to life as the desert valleys of Antarctica, are the most familiar examples of this survival talent—achieved through the symbiosis of algal and fungal components.

The land biota has done far more than just survive; the land biota now surpasses that of the sea in both biomass and diversity. Redwood trees, wasps, and other organisms visible to the unaided eye are actually habitats for rich ecosystems of smaller organisms that cover their surfaces, probe their interstices, take up residence in body cavities, and invade (sometimes peacefully, sometimes not) even the inner sanctum of their cells. Marine animals and even the few plants that live in marine environments (algae are not members of the kingdom Plantae; they are protoctists) also have symbionts and parasites, but the relationships seem sparse and platonic compared with the orgy on land.

Organisms, which are all primarily water, can interact at arm's length, so to speak, only in water. On land, direct physical connections become essential. Overall, terrestrial organisms had to build for themselves structures and components that could perform the environmental services that marine organisms can take for granted.

Yes, certain substances essential for life (oxygen and carbon, notably) are distributed more efficiently by atmosphere than ocean. But what about elements that never take on a gaseous form—like phosphorus? What about calcium and magnesium? What about any number of so-called trace minerals vital to animals and plants, be they marine or terrestrial? Compare, for example, the number of filter-feeding animals on land with that in the sea.

In a way, the land biota has had to find ways to carry the sea within it and, moreover, to construct watery conduits from "node" to "node." And to do so, this biota has evolved one entirely new kingdom of life whose importance is only now beginning to be understood: the fungi. Without the web of fungal hyphae in forest soils (their power masked by their delicacy), there would be no redwood trees; there would likely be no trees. The anastomosing (merging) habits of fungi and their unique enzymatic capabilities enable them to surround the surfaces of tiny particles of rock, extracting crucial nutrients, and then to probe or infest the tissues of tree rootlets to deliver the goods—in exchange, of course, for carbohydrates manufactured by the partner tree.

Land organisms have, by necessity, evolved together as part of a greater interconnected mass of living cells. In moving out of marine waters, complex life has taken the sea beyond the sea and folded it back inside of itself to form Hypersea.

The Hypersea perspective is thus an invitation to regard the land biota in a new way—as a remarkable interaction of cells that is far more than just an aggregation of organisms that succeeded independently in finding ways to live on land rather than in the sea. Hypersea encourages us to view the mass of terrestrial organisms as a geophysical phenomenon with unique properties of mass and velocity. If it is meaningful to speak of Lake Baikal, the Alps, the Ganges River, or the Arctic Ocean, then it is also meaningful to speak of Hypersea. Hypersea is more dispersed than the Alps, yet it is far more geochemically important—that is, it has a greater affect on the distribution and movement of molecules on Earth's surface.

In this book, we intend to demonstrate how the Hypersea perspective can be scientifically fruitful. It suggests new questions, answers old questions, and offers solutions to long-time enigmas. For example, a Hypersea perspective offers a new explanation for the genesis of coal. It also may explain why the land biota is so much more diverse than that of the sea, and why the land supports so incredibly much more biomass.

Hypersea unifies what might at first glance appear to be opposites. Plants and land animals are shown in our new view to belong to the same collective, and, geologically speaking, plants and land animals ought to be considered as part of a unified whole. Cases of parasitism and disease are related by Hypersea perspective to examples of mutualistic symbioses where the hosts and symbionts both benefit by living together. Natural selection, generally considered as a force of discrimination acting on separated individuals, evokes curious happenings on the insides of organisms.

The term *Hypersea* requires more care in its definition than usual for a new scientific term because of its multifaceted nature. In this book we use the term to mean three different but closely related things: a physical entity, a scientific theory, and a new way to view terrestrial life.

First, Hypersea is a physical—or more precisely, biogeo-physical—entity, composed of the bodies and fluids of all the land organisms with nucleus-bearing cells (plants, animals, fungi, and protoctists) and their symbionts. The boundaries of this living landform are complex but the main concentrations of Hypersea will be unmistakable as you read more about it.

Second, Hypersea is a new scientific theory. All useful theories are subject to falsification. Either the continents drifted or they stayed put. Either the organelles of cells originated as symbionts or they formed in some other way. Either the symbiotic interconnectedness of land organisms is the major factor responsible for the high diversity and vast biomass of the terrestrial biota, or something else is. Evaluation of this last statement constitutes the core of any research program that the Hypersea theory might inspire. If Hypersea theory survives a sufficient number of tests of its evolutionary and ecological significance, then it may eventually gain widespread acceptance in the scientific community.

Third, Hypersea is a novel perspective or scientific outlook on the land biota. We invite you to view land life through the Hypersea lens, and we believe that, after viewing it with this perspective, the living world will never look the same again. The Hypersea perspective will allow you to ask questions about land life that you may not otherwise have thought to ask. Hypersea as a perspective or scientific generalization is thus closely related to Hypersea the theory, for many of the questions this new perspective might generate could themselves be testable hypotheses with important theoretical implications.

Hypersea in the Fossil Record

The existence of Hypersea can first be seen in the fossil record of the early Devonian, 400 million years ago. Specifically, preserved in the Rhynie Chert is a record of intimate ecological interactions among plants, fungi, and microbes. The Rhynie Chert deposit, which occurs near Aberdeen, Scotland, is just older than the world's first major coal deposits.

The Rhynie Chert is without question one of the most important fossil deposits in the world, and it is the single most important site for Hypersea. The trenches from which the chert was excavated have since been filled with soil and grassed over, and the land is now rented out to a farmer to pasture his cows.[2] Too bad the cows cannot appreciate the hallowed ground over which they graze! The site has also been shown to be gold-bearing; apparently the gold was brought to the surface in the same hydrothermal, volcanic fluids that delivered the silica responsible for preserving the fossils when the silica solidified into chert.

Chert, also known as flint, is a smooth-feeling rock composed of submicroscopic quartz grains called cryptocrystalline quartz. Chert can form from solution, like rock candy, and it often traps organic matter, preserving it in recognizable condition for hundreds of millions of years. The Rhynie Chert was probably formed by ancient hot springs that acted as a source for silica-charged mineralizing fluids. The fossils within it are therefore of terrestrial and freshwater aquatic organisms.

Fossils entombed in chert are studied by thin-sectioning the rock. A slice of rock only a twentieth of a millimeter thick is mounted on a microscope slide with epoxy. Light can be transmitted through the thin section, illuminating for study microscopic fossils and microscopic relational features of macroscopic fossils.

The fossil record of fungi is an underappreciated source of information about the history of life. Thomas N. Taylor of Ohio State University and Winfried Remy and Hagen Hass of the Westfalische Wilhelms University in Germany described a fossil fungus inside a fossil alga (figures 1.2 and 1.3), which in turn occurs inside a fossil plant stem from the Rhynie Chert. The alga may have been parasitic on or mutualistic[3] with the plant, and the fungus was certainly parasitic on the alga because it belongs to a group whose living members are all obligate parasites. This sample of the Rhynie Chert thus provides the oldest known fossil of a parasite living within the body of

FIGURE 1.2 (Left)

Fossilized algal cells containing fungal (chytrid) spore chambers called zoosporangia. Both the algal cells and their internal fungal spore chambers occur *within* the stem of a 400-million-year-old plant. Maximum width of cell cluster is 52 microns.

FIGURE 1.3 (Right)

Detail of fossil spore chamber inside of an algal cell, both contained inside the stem of a fossil plant. Note discharge tube through which fungal (chytrid) spores exit the chamber. Greatest width of algal cell is 18 microns.

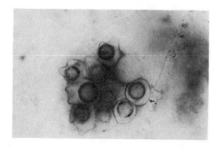
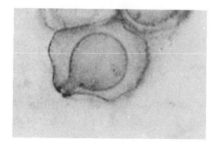

another (probable) parasite. A parasite within or on another parasite is called a hyperparasite.

We have recently made other fungal discoveries from the Rhynie Chert in our laboratory at Mount Holyoke College. These include specimens of fungi associated with fossil plants that are strikingly similar to the mycorrhizal fungi associated with modern plants. Mycorrhizal fungi live in intimate symbioses with the roots of plants and help those plants obtain nutrients from the soil; the fungi, in turn, receive sugars from the plants. In a number of cases, mycorrhizal fungi provide nutritive linkages between the roots of unrelated plant species. Figure 1.4 shows a cluster of fungal spores occurring as fossils in the Rhynie Chert from a pocket within a fossil plant. This specimen resembles the modern mycorrhizal fungus *Sclerocystis*. Other fossils in the Rhynie Chert look like the mycorrhizal fungus *Glomus*.

Both types of fungal fossils strongly suggest that mycorrhizal symbioses, with the fungi living partly or entirely within the body of the host plant, were beginning to be established by the Devonian. Furthermore, the most recent data indicate that fungal *hyper*parasites[4] were also present in the Rhynie Chert. Evidence for fossil hyperparasitism consists of fungal spores with hyperparasitic fungi on their surface or penetrating through the spore cuticles. A reaction zone or zone of host

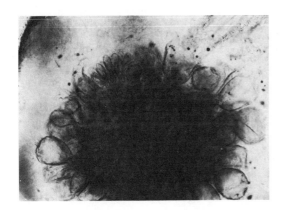

FIGURE 1.4

Fungal spore bodies resembling the modern mycorrhizal fungus *Sclerocystis*. These fungi lived within the tissues of a plant 400 million years ago, from the Lower Devonian Rhynie Chert, Scotland. Greatest width of specimen is 580 microns.

response to the invasion, identical to that seen in modern fungal symbioses, is seen in the Rhynie fossils.

The early Devonian is thus the minimum date for the origin of Hypersea—that is, for the initiation of the land biota's transition to an intimate association of cells networked for fluid circulation and matter transport. Judging from the Rhynie plant tissues, Hypersea was relatively well established at this time and so may have arisen a good deal earlier.

This brief survey of the earliest fossil evidence of hyperparasites in the land biota brings us to our first major prediction. *We predict that hyperparasites alive today will be found to be more common on land than in the sea.*

Internal Ecology: Hypersea in Animals

Our prediction that hyperparasites are more abundant on land applies not only to nested associations of plants and fungi but also to all sorts of intimate symbioses involving members of the animal kingdom. Parasitic animal symbionts come in two varieties: "regular" parasites, which do not usually kill their host, and parasitoids, which usually leave the host dead after completing the parasitic stage of their life cycle.

Parasitoid wasps, for example, insert their eggs into the bodies or eggs of other insects; the wasp larvae then develop inside the hapless host, feasting on the host's tissues and fluids. In an amazing example of what might be called "internal ecology," hyperparasitoid wasps infect other hyperparasitoid wasps which in turn infect hyperparasitoid wasps which (whew!) infect parasitoids of a free-living insect host. The host's fate is dependent on whether the hyperparasitoids kill each other (thus sparing the parasitoid and sacrificing the host) or whether the hyperparasitoids kill the parasitoid (thus saving the host if the kill occurs before the host is fatally injured). In another case, hyperparasitoid eggs are deposited on plant matter ingested by the host. Once ingested, the eggs hatch and then seek out and infect larger parasites already within the body of the host. These cases[5] indicate that the body of the host is essentially a self-contained habitat, complete with territorial interactions between parasites.

The accompanying table shows the contrasts in vocabulary used to depict the components of nested parasitism. The first column indicates the general host-parasite-hyperparasite distinction. The second column shows the traditional scheme used by agricultural entomologists. The third column shows the equivalent terminology of parasitologists. We prefer the H_x scheme shown in the fourth column (H expressly refers to Hypersea, but it can just as easily be thought of as a way to refer to hyperparasitic ecologies). H_1, for example, refers to

VOCABULARIES FOR NESTED PARASITISM

Traditional	Entomology	Parasitology	Hypersea	Example
Host	Host		H_1	Frog
Parasite	Primary parasite		H_2	*Zelleriella* (protoctist)
Hyperparasite	Secondary parasite	First-degree hyperparasite	H_3	*Entamoeba* (protoctist)
Hyperparasite	Tertiary parasite	Second-degree hyperparasite	H_4	*Sphaerita* (chytrid)
Hyperparasite	Fourth-stage parasite	Third-degree hyperparasite	H_5	
Hyperparasite	Fifth-stage parasite	Fourth-degree hyperparasite	H_6	

the host; H_2 to the parasite. H_3 and onward pertain to the hyperparasitic levels. The real-world example included in the table begins with a frog host, which contains a protoctist, which contains another protoctist, which contains a chytrid, (a funguslike organism). If the chytrid were to pick up a bacterial infection, this particular nesting would rise to the H_5 level. To our knowledge, the deepest nesting of hyperparasites occurs with a caterpillar serving as host and a parasitoid wasp at the H_6 level.

Another (though far less complex) example of nested parasitism can be found in your own home. If you try to get by in the kitchen with keeping flour in its original paper bag, sooner or later you will find yourself with unwelcome guests. At first you may not notice the oddity of a small moth happening to flit through your kitchen when you open a particular cabinet. But eventually you will notice the larvae—perhaps at a most inopportune moment. The larvae of the flour moth (*Plodia interpunctella*) are masters at coping with aridity; they need nothing more than dry flour (or oatmeal or pasta—nothing is safe!). These maggotlike creatures that may surprise us during baking may, however, be attacked by the parasitoid wasp *Venturia*. *Venturia* is not the only possible attacker, as other tiny wasps also favor moth larvae as a site to lay eggs. *Venturia* is, however, a very jealous parasite, and its larvae will attack other challengers for the host's internal resources with a threatening pair of sickle-shaped mandibles[6] (figure 1.5).

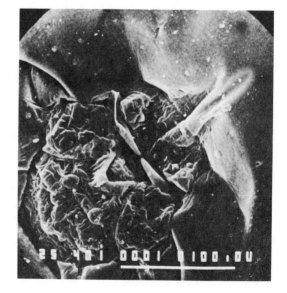

FIGURE 1.5

Electron photomicrograph of the mandibles of *Venturia canescans*. Larva is 180 hours old, and lived as a parasite within the body of a flour moth larva of the species *Plodia interpunctella*. Scale bar = 100 microns.

Even more bizarre examples of parasite competition within the bodies of host insects are known. After a parasitoid wasp belonging to the genus *Copidosoma* lays an egg in an egg or larva of its host (the moth *Trichoplusia*), the embryo hatches and, still within the body of the host, undergoes what is called polyembryonic reproduction. The embryo divides a number of times to form a sizable brood of identical siblings. A few of the embryos in the polyembryonic cluster undergo precocious development and begin to prematurely move away from the sibling embryo cluster and travel throughout the body of the host. Ultimately, the precocious siblings commit suicide by their early development because, although they grow to a relatively large size, they never mature into adult wasps. Instead, they cruise through the tissues of the host, seeking out and consuming nonsibling competitors. Competing parasites from other species are killed by the precocious warrior larvae, as are competing embryo clusters from unrelated *Copidosoma* broods. But it is the clonal sibs of the warriors, not the warriors themselves, who emerge as victors. Once the "normal" siblings mature, the warriors perish along with the spent host.[7]

These examples of nested parasitoids and competition among parasites lead us to regard the bodies of macroscopic terrestrial plants and animals as the setting for extremely active, if miniaturized, ecological interactions. The more we examine this phenomenon, the more convinced we are that such interactions are the rule, not the exception. These interactions constitute Hypersea. Hypersea interactions are pervasive, extending into every nook and cranny of complex life on land.

Human bodies are global outposts of Hypersea in the sense that they, too, offer lush internal ecosystems. Despite advances in health care during the last century, access to our own tissues by other organisms is not as restricted as we might like to think. As will be discussed in the final chapter, the microbe *Toxoplasma gondii* is present in virtually all populations of humans, with infection levels ranging from 15 to 85 percent of the adults.[8] As much as half of the population worldwide harbors the follicle mite *Demodex*. Infections of both parasites are typically asymptomatic. Multicellular internal parasites of humans, such as the dangerous schistosomiasis fluke, may be subject to hyperparasitism by *Toxoplasma* microbes. Recent evidence suggests that the microbe *Mycoplasma penetrans* is a cofactor, along with the HIV virus, in AIDS.[9]

Ticks (figures 1.6 and 1.7) are capable of transmitting to their mammalian hosts a wide variety of microbes. One bacterial pathogen transmitted by a hard tick (*Ixodes scapularis*) is the spirochete *Borrelia burgdorferi*, the microbe that causes Lyme disease. *B. burgdorferi* has a remarkable variety of tick and vertebrate hosts, and it can be passed from mother tick to offspring by a process known as transovarial transmission.[10] Another hard tick, *Ixodes ricinus*, which is known to transmit the Lyme disease pathogen to humans in Europe, is parasitized

FIGURE 1.6

Mammal-biting tick, ventral (belly) view. Specimen collected in South Hadley, Massachusetts. Scale bar = 1 mm.

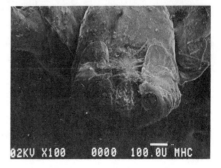

FIGURE 1.7

Detail of tick's biting mouth parts (same specimen as in figure 1.6). Scale bar = 100 microns.

by the wasp *Ixodiphagus caucurtei* (figure 1.8). This wasp may employ viruses to weaken the tick's immune system. Can microbes or viruses pass from the wasp, through the tick, and into the mammal? This is not known, but a Hypersea perspective leads us to predict that it may be more likely than one might think. For instance, a tick-associated illness resembling Lyme disease has been described in Australia, but the infectious agent has not yet been isolated from ticks, human patients, or parasites of ticks.[10]

As we hope to show, intimate interactions on land between completely unrelated organisms are exceedingly common, and they have led us to believe that Hypersea—the propensity of life on land to establish webs of physical intimacy and fluid exchange—is a dominant feature of the modern terrestrial world. Even our own mammalian tissues are very much a part of Hypersea.

Solving the Diversity Paradox

Hypersea can help resolve what we like to call the diversity paradox. It has long been recognized that there are many more

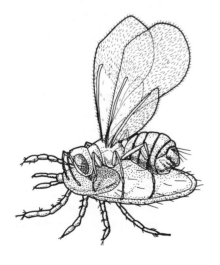

FIGURE 1.8

Ichneumon wasp *Ixodiphagus caucurtei* depositing eggs into a nymph of the tick *Ixodes ricinus*. *I. ricinus* is a European genus known to transmit Lyme disease to humans (the disease is caused by the spirochete bacterium *Borrelia burgdorferi*). It is the tick nymph that does the biting and transmits the bacterium. Whether microbes can be passed from ichneumon to tick to human is not known; however, the wasp may use a symbiotic virus to weaken the immune system of its tick host.

species of land organisms than marine organisms (this, despite recent evidence that benthic life in the deepest parts of the ocean is more diverse than previously assumed). We believe this discrepancy constitutes a paradox. Consider: even the most equable land environments must be harsher for life than is the sea; the most equable land environments (tropical rain forests) contain far more species than do tundra and desert; then why does land contain more species than ocean? Why does the land biota appear to be so diverse when compared to the marine biota? One would expect, based on the comparative physical harshness of the land environment, that the land diversity would be the lesser of the two.

In an early attempt to answer the question, scientists argued that the high diversity of terrestrial plants was responsible for much of insect diversity. Others have argued that the land environment fosters geographic isolation more than does the oceanic environment. This argument has some validity, if one considers land animal parasites restricted to a single host species. Geographic isolation on land, however, is in our opinion not the main cause of high land diversity, considering that many marine environments (lagoons, deep-marine hydrothermal vents, submarine caves) are also quite isolated. Further, marine creatures (both living and extinct) with nonswimming larvae are extremely evolutionarily isolated. And some marine parasites and hyperparasites are restricted to single host species.

More recently, some have contended that there are more species on land because the primary producers on land (such as trees) are quite large, so they can support more species than can be supported by the commonest primary producers in the sea (algae and bacteria). Others believe that the diversity paradox is myth and that the diversity of marine organisms is much greater than generally thought.

Regardless of the ultimate outcome of the global species count, the average species diversity *per unit area* of planetary surface is substantially greater on land than it is in the sea. (The land constitutes only about 30 percent of the surface of

Earth.) At a minimum, this discrepancy in diversity per unit area calls for an explanation. In our view the Hypersea concept can help resolve the diversity paradox.

Changes in marine species diversity versus Hypersea (land) species diversity through time are illustrated in figure 1.9. (The coarse temporal grid obscures the deep, but relatively brief drops in diversity owing to global mass extinctions so that we can see overall trends.) The diversity of Hypersea has exceeded that of the ocean for more than 300 million years. Hypersea diversity and marine diversity have converged some-what during the past 150 million years, during which time the continents have been in latitudinal positions unusually favor-able to the speciation of marine invertebrates by providing physical longitudinal barriers (long stretches of open ocean) that complement the thermal latitudinal barriers caused by the fact that it gets cooler as one moves toward the poles.

The crossover point in figure 1.9, when the diversity in

FIGURE 1.9
Logarithmic comparison of estimated marine and hypermarine diversity over the last half billion years of geologic time. (Mass extinctions are too brief to show on this graph.) We estimate that hypermarine diversity has exceeded that of marine diversity for the last 350 million years.

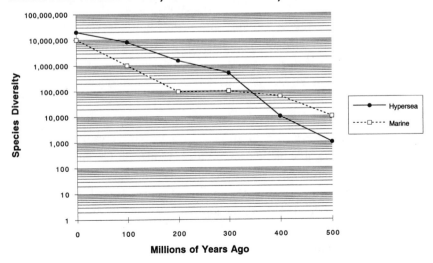

Hypersea overtook marine diversity, occurred during the Mississippian (or earliest Carboniferous), approximately 350 million years ago. Note that this crossover is the same age as the earliest coal deposits. This is an important point that will be covered in chapter 7. The exact time of the crossover may or may not indicate a special event—but surely the steeper slope of the curve compared with that of the marine demands an explanation. Our explanation focuses on hyperparasites.

What we call Hypersea, remember, formed as the result of direct physical connections between organisms striving to conserve water in the desiccating terrestrial environment. Nutrients on land came to be distributed among unrelated organisms via direct physical connections in the form of mutualistic, parasitic, and hyperparasitic interactions. If our Hypersea-derived prediction that land hyperparasitism today exceeds that of the sea is correct, it may be because hyperparasitism is a particularly successful strategy for the exploitation of the Hypersea environment.

Not only should the actual numbers of hyperparasite species on land exceed those of the marine realm, but hyperparasites should constitute a greater proportion of diversity on land than they do in the sea. One could attempt to count the hyperparasites directly to test this prediction. But a difference may show just as well by looking only at the highest levels of hyperparasitism—what we have called H_4, H_5, and H_6. Specifically, *we predict that, for at least the last 350 million years, the land biota has had several more higher levels of hyperparasitism than the marine biota has had.*

Hypersea can also help explain why biotic communities on land are richer in species on a per-unit-area basis than are comparable biotic communities in the sea. (The species richness on land is especially striking if one considers the low ecological efficiency—that is, the amount of energy passed from one level of the food chain to the next—of the land biota compared with the marine biota.[11]) Our predicted differences in number, nestedness, and overall importance of hyperparasites between land and sea would certainly contribute to a higher average diversity of land organisms per unit area. More important, however, is the fact that on land so many species of organisms are capable of thriving *in* the body fluids of other organisms. Additional species are literally packed inside of or on top of one another. In an example from Panama, there is a one-to-one correspondence between eleven species of fig, their eleven species of pollinating wasps, and the eleven different nematode species that are internal parasites of the wasps. The relationships are so specific that the nematodes have species names derived from the name of the fig tree in which their host wasp lives.[12]

Other insects harbor multiple types of nematodes within the same species. The bark beetle (genus *Ips*) is parasitized simultaneously by two different genera of nematodes.[13] Such situations are not uncommon in the tropics. An herbivorous insect may be attacked by up to ten species of parasitoids.

The large number of cases like this have led us to view the colonization of the land by "advanced" life as the colonization of two environments: (1) the barren land surface itself, and (2) the Hypersea environment formed by the tissues and body fluids of plants, animals, and fungi that themselves colonized the land.

This "colonization within a colonization" may explain why there are approximately two orders of magnitude difference between the modern land biomass and the modern marine biomass. The usual explanation for this disparity is that land plants store a lot of this biomass as "dead" wood. Much of the wood, however, is actively involved either in transport of fluid to adjoining live tissue or in the nourishment of fungal mycelia. Hypersea host species actively create, rather than passively inhabit, a fluid medium.

The Hypersea perspective also explains why, as traditionally tallied, marine food chains are long and land food chains are short—land food chains are internalized to such an extent that they are hidden from the casual observer or even from the careful observer with the wrong perspective. Land food chains also have lower ecological efficiency than do marine food chains. This means that more energy is lost in going from, say, a gazelle (prey) to a cheetah (predator) than from a mackerel (prey) to a tuna (predator). A significant fraction of the seemingly wasted energy is consumed by organisms hidden within their Hypersea hosts.

Hypersea is manifest as a *directed* transfer of fluids. This function involves the transfer of matter in both horizontal and vertical directions. The physical environment of the oceans, of course, produces horizontal transport with very little help from the organisms themselves. The land biota must be far more involved in lateral movements of fluids and materials. In the oceans, vertical upwelling happens only (but mightily) in certain locales, and the rising nutrients are then distributed by currents. Marine organisms are (with one interesting exception) passively dependent on currents and circulation patterns to supply nutrients. But the fluid-lifting abilities of land organisms allow "upwelling on demand." Nutrients from Earth's crust (or in the case of nitrogen, from air diffused into the soil) are continually being captured and carried to the surface by plants.

Plankton and other organisms growing and reproducing at the sea surface overconsume their local marine nutrients to the point of famine. Ocean surface nutrients can only be replenished by nonbiological forces such as wind mixing or by seasonally or geographically restricted upwelling. Marine upwelling is very restricted in extent and is often seasonal even in the tropics. Hypersea organisms, in contrast, effectively bypass what could otherwise have become a similar nutrient starvation on land.

Hypermarine "upwelling" is a biotic affair, involving the mycorrhizal fungi and their interconnected vascular plants. We believe this is why the primary productivity of the land so outstrips that of the sea. Nutrient limitation, which drastically curtails productivity in surface marine waters, is not a big problem on land. During the active growing season, the land biota creates for itself a zone of constant upwelling. (Later in the book we will argue that hypermarine upwelling, measured as a geochemical force, is comparable in scale to both photosynthesis and the limestone secretion by marine organisms.)

In summary, the Hypersea concept confirms the sense among ecologists that fundamental differences exist between terrestrial and marine ecosystems. But we believe we know the difference that makes the difference (as Gregory Bateson would say), and that difference is the evolution of techniques for directing nutrient flow in the land biota by the land biota.

TWO

The Interior Sea

that marvelous sap   —*Goethe, in reference to blood*

Water is sometimes called the universal solvent because it exists on Earth in predominantly liquid form and because it has one very special property. Water dissolves more different types of compounds, and in greater quantities, than does any other common liquid. Water is also an inert solvent, meaning it does not chemically change under the effect of the substances it dissolves. This trait is invaluable for living organisms; it allows an organism to use a single water molecule repeatedly. As a chemical catalyst, water never wears out.[1]

The actual amount of water involved in the growth and maintenance of the land biota is minuscule when compared to the earth's total surface water supply. At any instant, the modern land biota utilizes only 1.5 billionth (1.5 parts in 1,000,000,000) of the total volume of water on Earth.[2] Yet this tiny amount of water has an enormous impact of the geochemistry of the planetary surface, not to mention its physiological significance for the land organisms themselves.

The biomass of the continental biota exceeds that of the world's oceanic biota by several orders of magnitude, with the biomass of forests making up 84 percent of the terrestrial biomass (total weight of living material).[3] These facts beg the question: If the oceans cover two-thirds of the earth's surface, and water is the most essential chemical component of life, why do the oceans contain only a few percent of Earth's biomass? This disparity is all the more remarkable when one considers that the land organisms work under the handicap of not being continuously bathed by water. They need to episodically replenish their water supply.

Blood as Seawater?

In 1957 the celebrated film director Frank Capra created an educational film entitled *Hemo the Magnificent.*[4] This film served for many years as an engaging introduction to human physiology, with special emphasis on the circulatory system. With the help of an arrogant Grecian cartoon character, "Hemo," the audience was given a fairly sophisticated yet thoroughly enjoyable introduction to the feedback loops involved in controlling the operation of the human circulatory system. The film portrayed the saltiness of blood as a legacy of the marine environment of our fishy ancestors. Capra got a great deal of mileage out of the appealing image that the nurturing environment once outside the bodies of our ancestors is now within, coursing through our arteries.

The film mentioned that tropical animals "might exist" in the 1 percent salt content of blood plasma, and at one point Hemo announced that "in the beginning," blood was the sea and that blood was and is "a specialized form of seawater." At the time of the film's production, there was only limited support in the scientific literature for Hemo's assertions.

The idea of blood as seawater was scientifically developed by A. B. Macallum in the 1920s. Macallum maintained that by studying the blood of freshwater fish one could determine the solute composition of middle Paleozoic seas. He argued that freshwater fish blood, somewhat less salty than modern seawater, indicated that the world ocean had once been less saline. Macallum's ideas gained wide and fairly long-lasting acceptance. A skeptical essay was published in 1931 regarding Macallum's claims, but it was not until the early 1940s that the tide of scientific opinion turned against Macallum. The fatal fact for the hypothesis was the demonstration that at least for the last 700 million years there has been little change in the proportions of the major ions of marine water. Good evidence now exists that the chemical composition of the oceans has stayed pretty much constant for a long time, and that the differences between seawater and cell fluids or body fluids are not related to the history of ocean chemistry but to the various biochemical reactions in which ions have a function. The use of oxygen isotope variations to determine ancient seawater temperature, a proven technique, assumes constant marine salinity for hundreds of millions of years.

Capra's tale about our blood being the remnants of the old sea that gave birth to us vertebrates was lovely, and it may have helped a lot of people bond with the planet and enjoy the idea of evolution, but it was a fairy tale. Vertebrate blood is not only less salty than the sea today, it has likely been less salty ever since the origin of freshwater fish.[5]

The "blood as seawater" hypothesis was abandoned by serious scientists, even while it was retained as a popular heuristic image in *Hemo the Magnificent*. The image of blood as seawater is intuitively appealing, but it cannot be taken at face value. Nevertheless, there may be some validity to Macallum's claim. Even if not applicable with regard to the transition from sea to land, it may be valid for the reverse. Vertebrates such as cetaceans and sea turtles that bade goodbye to the land and their Hypersea companions have blood which is nearly always more dilute than the marine surroundings (the exceptions being estuaries flushed by great rivers). These animals must take physiological steps to avoid water loss through the osmotic pull of their salty surroundings. The dilute character of the blood of vertebrates that have returned to the sea indicates an evolutionary history in or near fresh water.

In this book we intend to resurrect, albeit in a modified form, the hypothesis that the sap of plants and the blood of land animals has an evolutionary connection with seawater. Living fluids are not a mere remnant or analog of the sea; they are actually a *new* type of sea or marine environment: Hypersea.

The Rising Tide of Land Life

The biogeochemical and evolutionary importance of the land biota has, in our view, been hugely underestimated. We believe that the appearance of complex life on land was a major event in which a kind of mutant sea invaded the land surface. It was as if the nimble offspring of the old sea had learned how to slosh and slop up onto land, with the tissues and vascular systems of land organisms acting as a complex, water-retaining sponge. Cuticle and skin took the functional place of the surface tension of water where sea meets air.

The land biota represents not simply *life* from the sea, but a variation of the sea itself. Acting over evolutionary time as a rising tide, the land biota literally carries the sea and its distinctive solutes over the surface of the land, into some of the driest environments on Earth.

But traditional study of the emergence of land life has missed the grandeur of the event from a geochemical perspective and has focused too intently on detailed differences between related terrestrial lineages. We prefer an approach that considers the entire ecosystem as a unit and that analyzes the relationships among the various biotic subunits. This approach to an understanding of terrestrial organisms can help weed out old falsities born of conceptual oversimplification. As an illustration, consider the terrestrial fungi.

Terrestrial fungi are commonly portrayed as mere saprotrophs, forms that feed on dead organic matter. But this is a misleading simplification. Fungi also parasitize animals and vascular plants (not to mention other fungi). The internal parasite approach may have been the only method by which fungi were able to leave the aquatic habitats that had served as the first habitat for this kingdom. Parasitism is likely to be the original fungal lifestyle on land; saprotrophic lifestyles were derivative.

Fungi, moreover, have become the world's expert parasites. And fungal parasitism is a key to understanding the colonization of the new land ecosystem. Of particular interest is the way many kinds of parasitic fungi use more than one kind of animal or vascular plant as vectors, or vehicles for completing different parts of a single, but complicated life cycle. According to plant biologist Gail L. Schumann:

Humans tend to worship what they don't understand, as illustrated by both rust and mistletoe (there are actually many kinds). The rust gods, Robigo and Robigus, were worshipped through the sacrifice of red dogs, cows, etc. in hopes of protecting grain from rust fungi such as *Puccinia graminis*. Since many Roman gods are similar to Greek gods, it's interesting that the rust gods are not found in Greek mythology. *Puccinia graminis* originated in the fertile crescent and has been found in archaeological ruins in Israel dated at 3500 B.C. but rust disease probably didn't make it to Italy until the alternate host, barberry, made its way to Europe with increased travel from the East. This fungus is amazing—requiring two unrelated plants to complete its life cycle. Unfortunately, both plants were brought to the western hemisphere by Europeans—but that's a whole story in itself. The myth goes that a bad boy set a fox on fire and let it run through the grain fields which angered the gods, so they punished people with rust disease. The story gives us the black and red imagery associated with the red and black spore stages of the fungus found on wheat. The festival was called the Robigalia and was eventually absorbed into the Christian calendar as Rogation Day, blessing of the crops, April 25. Mistletoe is an evergreen, parasitic plant found on hardwood trees, which also has long ties to Greek, Roman, and

Celtic mythology—an evergreen in winter that never needs to touch the ground! It was harvested only with golden sickles by druid priests. Mistletoe was thought to be the result of Thor's lightning bolts. The Christmas custom of kissing under mistletoe comes from the old tradition of enemies embracing to make peace under mistletoe at the winter solstice.

St. Anthony is saint for the victims of ergotism, the holy fire of the middle ages caused by the fungal ergot disease of rye and other grains. Ergotism perhaps contributed to the start of the French Revolution and also may have been involved in the accusations of witchcraft in Salem and elsewhere. The disease causes constriction of blood vessels leading to gangrene and loss of hands and feet.

Schumann's book *Plant Diseases: Their Biological and Social Impact* (1991, APS Press) is filled with other fascinating examples—such as an explanation for why the English have tea time rather than coffee breaks. Ceylonese (Sri Lankan) coffee plantations owned by the British were destroyed by coffee rust fungus, and the failed coffee crops were replaced by tea plantings.

In addition to alternating between hosts, parasites can transfer fluids from one host to the next. The body fluids of the hosts, in our view, become the waters of Hypersea, droplet by droplet. The organisms capable of both navigating and extending this new sea found the door wide open to strikingly new routes of evolutionary change.

Hypersea's Impact on Climate

Living things have influenced the composition of Earth's atmosphere for billions of years. The first organisms on land were probably the first organisms to seriously affect the atmosphere.

The first organisms on land were bacteria known as thermophiles. Thermophiles, or "heat lovers," prefer a hot, carbon dioxide–rich atmosphere, which is exactly what they found when they colonized the land. This initial colonization probably took place soon after the origin of life itself, sometime between 3.5 and 3.8 billion years ago, when surface temperatures likely averaged 70–100°C (160–230°F), and when land surfaces midday would have been far higher. The global average surface temperature today is only 15°C (59°F).

Thermophilic bacteria spreading over the dry planetary surface enhanced the weathering rate of rocks and minerals in early soils.[6] The increase in weathering meant faster breakdown of rocks into their constituent minerals, and faster chemical recombination of those minerals with carbon dioxide in the atmosphere.[7] In other words, the thermophiles had the inadvertent effect of drawing carbon dioxide out of the atmosphere and, by doing so, cooling the climate to their eventual detriment. By supplementing ongoing abiotic processes of chemical weathering, microbial colonization of the land, before the evolution of Hypersea, may have moderated the climate of Earth by as much as 35°C (63°F). Hypersea may have been responsible for the greater part of the later decline, down to today's global average temperature.

The carbon dioxide content of the atmosphere is modulated by the weathering of exposed silicate rock, which includes all igneous rocks and all sedimentary and metamorphic rocks that are not primarily limestones or other carbonates. This weathering, and the subsequent water-moderated recombination of constituent minerals by formation of marine limestones, is the primary path by which the planet offsets the continual outgassing of carbon dioxide through volcanoes and deep-sea vents, and it is the primary mechanism by which the planet achieved a net reduction in the atmospheric reservoir of carbon through geologic time.

It was the enormous amount of carbon dioxide in the early atmosphere of Earth that was making the planet searingly hot, as carbon dioxide is a greenhouse gas. Thermophilic bacteria obviously cooled Earth's surface too much for their own good.

Today thermophiles are restricted to rare, hot environments such as geysers and hot springs.

David Schwartzman, Tyler Volk, and I (Mark) have theorized that the progressive cooling of Earth's early atmosphere (largely caused by the biotic enhancement of weathering) enabled the evolution of more complex forms of life, whose proteins are otherwise unstable at high temperatures. So although thermophilic bacteria put themselves out of business by their enhancement of mineral weathering and the concomitant atmospheric cooling, they literally prepared the ground and air for the rest of us. There may be something about high ambient temperature that inhibits the evolution of complex life, for instance by preventing the creation of regulatory enzymes.[8]

The next big change in the biotic enhancement of weathering came with the appearance of land plants—with the origin of Hypersea. A casual glance around natural areas anywhere on land suggests that plants are the dominant terrestrial organisms. From a planetary perspective they are also dominant; viewed from space, Earth usually shows visual evidence of no other type of life. Land plants account for about 98 percent of Earth's standing biomass. Land plants contain, globally, between 420 and 830 gigatons of carbon. Soil organic material (attributable largely to plants) and leaf litter store another 1,200 to 1,600 gigatons.[9] The total biotic reservoir of carbon on land is huge and can be regarded as a storehouse of hydrocarbons and associated organic compounds.

As the global range of vegetation expanded in Permian and later times, plants colonized successively drier environments. By the late Mesozoic, much of the dry land surface of the planet was probably inhabited by plants. Hypersea had reached the full extent of its range, if not yet its richness.

Several researchers have tried to specifically link the evolution of land plants with the further enhancement of weathering of silicate rock (beyond what the microbes had earlier achieved) and thus the further diminishment of atmospheric carbon dioxide.[10] Robert A. Berner and Tyler Volk theorize that the rise and spread of land plants was a major contributor to an apparent increase in the weathering rate of silicate rocks—by a factor of about seven. Plants accomplish this by secreting acids and by recycling and holding moisture in the soil. The Hypersea community is better equipped to enhance weathering than are purely microbial communities. The fractal-like richness of Hypersea roots and hyphae create deeper and moister soils than microbial crusts, sitting on the surface of the ground, could ever achieve.

Some scientists have suggested that the diversification of angiosperms (flowering plants) in the late Mesozoic may have contributed to a pronounced increase in silicate weathering greater than that brought about by the more ancient gymnosperms (which include pines, conifers, and cycads). Other scientists dispute this claim, noting that if data from limy soils are excluded, then the supposed weathering rate difference between angiosperm- and gymnosperm-dominated ecosystems vanishes. Another objection to viewing flowering plants as extraordinarily good enhancers of rock weathering is that the largest increase in plant-related weathering rates in silicate rocks seems to have occurred during the Carboniferous, the so-called Coal Age. This is the time when terrestrial ecosystems really began to flourish—that is, when they expanded beyond the floodplains and coastal plains up into the highlands. Flowering plants did not arrive on the scene until over a hundred million years later.

The most profound geochemical impact of land plants was thus probably felt in the late Paleozoic, well before the appearance of flowering plants. The atmospheric carbon dioxide concentration was, according to Berner, considerably higher during the early Paleozoic than it is today. Carbon dioxide content of the atmosphere, Berner theorizes, was drawn down during the Paleozoic by the expansion and success of vascular plants. This drawdown occurred for two reasons.

First, as already noted, plants increased the weathering of silicate rocks. Second, vascular plants created a new source of bacterially resistant organic matter when they invented tough substances like wood cellulose and lignin for structural support. These substances tend to resist decomposition and thus provide the hydrocarbons in organic-rich sediments like coal. When these hydrocarbons were preserved by burial they were taken "out of atmospheric circulation," and the carbon dioxide content of the atmosphere fell accordingly. Berner notes that the global spread of woody plants must have been almost immediately followed by the spread of wood-decomposing fungi and bacteria. Otherwise, the power of plants to sequester carbon would have soon (geologically speaking) driven down the atmospheric content of carbon dioxide to levels so low that the plants themselves could no longer thrive.

Because of these two processes (biotic enhancement of silicate weathering and storage of carbon as coal), Berner suggests that plants caused the severe late Paleozoic glaciation called the Gondwanan Ice Age. If Berner is right—if the most extensive glaciation of the last half billion years, the Permo-Carboniferous glaciation or Gondwanan Ice Age, was caused by the drop in carbon dioxide as a result of the first big development of the land flora—then there can be no doubt that the primary constituents of Hypersea, terrestrial plants, have had a significant effect on atmospheric composition and global climate.

Hypersea's Impact on Evolution

The effect of plants on the atmosphere, in turn, has had a telling influence on plant evolution. As land plants began to flourish worldwide, carbon dioxide was pulled from the atmosphere. When carbon dioxide becomes scarce, the water-use efficiency of plants falls. They must open their pores wider and for longer periods in order to take in enough carbon dioxide to enable photosynthesis, and when they open their pores, more of their internal moisture is exposed to evaporation. When a plant experiences a lower level of carbon dioxide than that to which it is adapted, it may therefore be stressed both by starvation and by desiccation.

There was a fivefold decrease in atmospheric carbon dioxide during the late Devonian and Carboniferous. This late Paleozoic drop in atmospheric carbon dioxide has been invoked to explain increases in the diameter of the water-conducting cells in plants (tracheids) and certain modifications in leaves and roots that occurred in the Silurian and Devonian. For example, it has been suggested that the frequency of stomates in fossil leaf cuticle reflects ancient levels of atmospheric carbon dioxide.[11] These changes in plant anatomy ultimately proved to be springboards for future plant evolution.

Anatomical adaptations to carbon dioxide starvation also prepared plants to survive better under conditions of drought and cold. Plants therefore may have been very fortunate, because the same anatomical changes that are theorized to have helped them cope with lower levels of atmospheric carbon dioxide (which plants themselves may have caused) also helped plants survive times of global cooling (which plants may also have caused by lowering levels of atmospheric carbon dioxide). Plants were thus very well situated to expand their ranges during "better times," such as when atmospheric carbon dioxide and global temperatures returned to higher livels during the Mesozoic.

Plant activity has also been implicated as a controlling factor in the supply of nutrients from land to sea. Theories have been proposed suggesting that plants played an indirect role in, of all things, marine animal evolution. Some scientists claim that land plants accelerated marine animal evolution; others claim that land plants delayed or halted marine evolution. Both sides base their conclusions, however, on a shared belief that the rate of marine animal evolution is strongly affected by the rate of nutrient delivery to estuaries and shallow marine waters by river systems. This rate is controlled by the activities of plants on land.

In the mid Paleozoic, the delivery of nutrients to seawater

may have been increased by plants. The food-rich coal swamps of the Devonian, and the riverborne, seaward-flowing nutrients (in the form of carbon-rich debris) derived from early vascular plant floras may have helped trigger the evolution of the earliest modern groups of fish. The Devonian estuaries and tidal swamps (functionally similar to today's mangrove ecosystems), were probably the nurturing grounds for a significant evolutionary event that led to the origin of all modern groups of fish.

Subsequently, late Paleozoic plants expanded into upland environments and, ironically, may have caused a delay in the delivery of some nutrients to the sea even as the new upland plants were increasing the rate of rock weathering and nutrient release in these upland areas. Drops in the rate of delivery of phosphorus to the sea from rivers, perhaps a result of upland vegetation building soils and thereby holding on to the fruits of weathering, may have slowed the rate of evolution of highly cephalized marine creatures in the Permian and Mesozoic. No longer would the marine biosphere have exclusive claim to critical nutrients (such as phosphorus) derived from weathering.

Hypersea's Impact on the Oceans

As just discussed, various arguments have been presented implicating the presence of land plants as factors increasing or decreasing the amount of land-derived nutrients reaching the sea. Whether these putative changes actually occurred is not yet known, and the idea presents an inviting area for future research. It seems quite likely, however, that terrestrial life alters rates of erosion and weathering. It has even been speculated that, because of the tendency of plant roots to hold sediments in place and slow erosion, the land might be higher today than it was before 400 million years ago—that is, the slow gains made by ongoing tectonic uplift in certain regions are not so quickly lost to erosion and weathering when protected and reinforced by plants.[12]

On the other hand, plant-enhanced weathering of rock is between three and ten times faster than the weathering rate of rock colonized only by microbes. (Six is a reasonable approximation.) In other words, weathering occurs up to an order of magnitude faster when land plants appear on the land surface, even though the rate of delivery of sediments and nutrients to the oceans may be slower. Soil nutrients are thriftily used and reused by the land biota, especially when these nutrients are in short supply. Tropical rain forests receive up to twenty feet of rain water annually, and one might expect all nutrients and all minerals except the most insoluble iron oxides to be leached out of the soil by such a high water flux. Indeed this is the case, and tropical soils (called laterites) are stained red from the iron oxides. Yet the abundance of life in tropical rain forests demonstrates that the forests efficiently recycle needed nutrients to form an internal environment, a Hypersea, protected from the diluting influence of excessive precipitation.

Thomas J. Algeo and J. Barry Maynard have suggested that the middle Paleozoic episode of oceanic anoxia—an interval when the oxygen content of ocean bottom water plummeted— was caused by the first flourishing of land plants. The marine anoxia caused the deposition of organic-rich, black laminated shales. The existence of the shale, in fact, is how we know the anoxic event occurred. The episode of ocean anoxia probably was a significant contributor to the mass extinction of the marine reef-forming organisms of the middle Paleozoic. The anoxia can be attributed, at least in part, to a spurt in the oxygen consumption of bacteria that had to decompose huge volumes of plant litter delivered to ocean waters by streams. Basically, far more organic material was entering the oceans than the oceans had the capacity to oxidize; oxygen could not diffuse into seawater fast enough to replenish the oxygen consumed during decomposition. Once the oxygen was used up, additional increments of organic sediments could no longer be decomposed, and they became the black shales.

Activities of Hypersea on land affected oceans in another

way. Recent research[13] indicates that changes in clay mineralogy induced by early plants led to changes in ocean chemistry.
The clay minerals smectite and kaolinite both indicate deep
weathering in soil. The proportions of smectite and kaolinite,
in comparison to other clay minerals (such as chlorite and
illite), increase across the Devono-Carboniferous boundary,
strongly suggesting that the presence of plants altered the clay
mineralogy of the soil. When smectite forms, it absorbs more
magnesium than do other kinds of clay minerals, thereby lock-
ing magnesium into the soil and preventing it from reaching
the ocean. This effect may explain why the deposition of marine
dolomite (a magnesium-rich carbonate similar to limestone)
drops and the deposition of limestone (with its low magnesium
content) increases in concert with the expansion of vascular
plants in the Carboniferous. In other words, land plants caused
major changes in soil mineralogy, which in turn altered the
types of minerals deposited on the ocean floor.

THREE

The Marine Prelude

The ocean was the first place where organisms attempted to live on and in one another. Such symbiotic interactions are still important in marine ecologies, but predation and other forms of heterotrophy (consuming other creatures) are today the most typical ways for marine organisms to interact. Marine symbiosis has a long and important history, which we must explore in order to appreciate the veritable explosion of symbiotic interactions that makes Hypersea so distinct.

Symbiogenesis and the Origins of Complex Cells

By now most scientists involved in biology or paleontology accept the once-revolutionary idea that the origin of the cells of animals, plants, and fungi is linked to the symbiotic union of two or more types of bacteria. These more complex forms of cells are eukaryotes; they contain organelles—the descendants of once-independent bacteria—and, unlike bacteria, their genes are sequestered from the rest of the cell within a membrane-bounded nucleus. The four eukaryotic kingdoms of life are the protoctists (including unicellular forms and slime molds), plants, animals, and fungi. The fifth kingdom of life comprises the monerans, also known as prokaryotes or bacteria.

The idea that complex cells evolved through the mergers of simpler cells is called symbiogenesis: genesis through symbiosis. It has far-reaching implications, but the theory languished for decades among western scientists. It was viewed as fantastical and thus attracted little research. Russian scientists, however, had a well-developed research program dealing with this question by the end of the nineteenth century; they made great strides in the early part of this century.

Russian scientists Konstantin S. Merezhkovsky (1855–1921) and Boris M. Kozo-Polyansky (1890–1957) both correctly inferred that chloroplasts were derived from once free-living photosynthetic bacteria. Both men were influential scientists in Czarist and then Soviet Russia, but neither saw their ideas about the symbiotic origin of complex cells make much headway in the western scientific world. Indeed, it was not until the 1960s and 1970s, when Lynn Margulis conducted her research on symbiogenesis, that such ideas gained any credence at all in western Europe and the United States.

Merezhkovsky coined the term *symbiogenesis* in 1919. He defined symbiogenesis as "the origin of organisms through combination and unification of two or many beings, entering into symbiosis." Merezhkovsky saw symbiosis as a major source of evolutionary novelty, and he even held it responsible for the greatest innovation in the history of life since life's beginning—the origin of the eukaryotic cell.[1]

The fossil record is extremely unlikely to ever provide direct fossil evidence of the earliest eukaryotic cells. No marine fossils resembling eukaryotes are known until after the Oxygen Revolution of approximately two billion years ago. Nevertheless, the origin of these cells is quite evident in the clues given by living organisms today.

The Oxygen Revolution

The earliest organisms were all anaerobes; they lived in the virtual absence of free oxygen. For those earliest prokaryotes, free oxygen would have been toxic. When one lineage of prokaryotes, the cyanobacteria, evolved a way of capturing the energy of sunlight through photosynthesis, the clock began ticking for the day when oxygen, the waste-gas they produced, would effectively poison the planet. But evolution has a marvelous way of finding opportunity in disaster. Aerobic forms of life evolved that not only found ways to detoxify oxygen with special enzymes, but also found ways to use it to their advantage. Respiration of oxygen made it possible for an aerobe to produce far more useful energy than could an anaerobe.

The Oxygen Revolution is that moment in geologic time when significant quantities of free oxygen first accumulated in the atmosphere and the sea. Before then, oxygen was absorbed and rendered harmless by chemical combination with marine iron and other substances common in seawater. But these oxygen sinks eventually reached the limit of their ability to chemically combine with oxygen, and continuing photosynthesis by cyanobacteria meant that the waste gas had "no place to go." Oxygen began to accumulate in the atmosphere.

In the earliest part of Earth history, before the development of an oxygenated atmosphere, native iron could exist at the surface of the earth without rusting. Iron would however dissolve, because unoxidized (native) iron is soluble in water. This is exactly what happened; early streams and rivers carried into the sea tons upon tons of dissolved iron (derived from rock weathering). So for hundreds of millions of years, whenever a molecule of oxygen was given off by a cyanobacterium or photosynthetic protoctist, that oxygen immediately combined with marine iron.

The oxidation of marine iron was a two-step process over time. In the first stage, oxygen combined with native iron to form ferrous iron (FeO). Ferrous iron remained soluble in water. Once the marine iron had been oxidized to this level, the second stage began. In the presence of more oxygen, ferrous iron combined with oxygen to form ferric iron (Fe_2O_3), which is fully oxidized. Ferric iron, however, is insoluble; it precipitated and sank to the bottom of the sea where it lithified into the geologically famous (and never to be repeated) banded iron formations, which today are the primary commercial sources of iron ore. The Oxygen Revolution caused the oceans literally to rust.

The oldest fossilized eukaryotic cells are dated to about the time of the Oxygen Revolution. In the Gunflint Iron Formation of the Great Lakes Region there occurs a microfossil called *Eosphaera tyleri*[2] (figure 3.1). This tiny fossil, little more than ten microns in diameter, consists of an inner spherical wall and outer spherical wall. Between the two walls are smaller, blisterlike spheres. The smaller spheres may represent endosymbionts—that is, small bacteria living within a larger host, perhaps living there because the host had the special enzymes (such as superoxide dismutase) used to detoxify oxygen. One can scarcely imagine a more compelling incentive than oxygen poisoning for small bacteria to cohabit within the body of a larger host that had evolved a protective enzyme system.

The oldest fossils that can be firmly interpreted as protoctists (and therefore, eukaryotes) are acritarchs 40 to 200 microns across, from China. Acritarchs are hollow spheroidal microfossils with resistant organic (sporopollenin) walls. They are thought to be the resting cysts of marine photosynthetic plankton. Acritarchs are separated from rock for study by dissolving the silicate rock matrix with hydrofluoric acid. The Chinese specimens are approximately 1.8 to 1.9 billion years old and are the oldest widely accepted fossil eukaryotes.

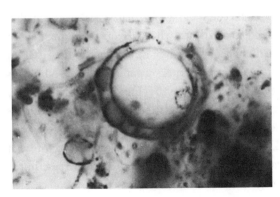

FIGURE 3.1

Eosphaera tyleri, a two-billion-year-old fossil microbe from the Gunflint Iron Formation, Canada. This microbe may have been one of the earliest eukaryotes. Width of the outer wall is approximately 10 microns.

Carbonaceous megafossils ("mega" meaning big enough to be visible to the naked eye) are presumed to include remains of eukaryotic multicellular algae. Carbonaceous megafossils first appear in the early Proterozoic about 2.1 billion years ago. These fossils, shaped like tiny pancakes or flattened sausages, are considered to be eukaryotes because they are carbonized impressions of organisms that were greater, and often much greater, than about sixty microns in diameter. Paleontologists have tried to infer the date of the origin of eukaryotes by looking at the sizes of fossilized cells. Those smaller than ten microns in diameter are assumed to be prokaryotes. The problem with this assumption is that, in fact, eukaryotes as small as one micron in diameter are known to exist today.

All eukaryotes are obligate aerobes; that is, they cannot survive without oxygen. A reasonable case can thus be made from the fossil evidence for the origin of eukaryotes any time after the oxygen crisis. A plausible case can even be made for the origin of tiny eukaryotes (indistinguishable from prokaryotes when fossilized) before then; perhaps the earliest eukaryotes lived in close association with cyanobacteria in order to capture oxygen waste gas before it had a chance to encounter the oxygen-loving solutes of the sea.

Who Were the Symbionts?

The creation of eukaryotes by symbiogenesis is an event that may have occurred many times when various combinations of microbes in earth history found it advantageous at different times to fuse into a single, larger organism. Lynn Margulis considers symbiogenesis the main driving force in evolutionary change. In some cases, the new relationship eventually becomes an obligate one—and when this happens, a more complex cell has evolved. This kind of symbiogenesis is an example of what the Russians call arogenesis, or the evolutionary increase in the complexity of organisms.

Only a few of the many possible combinations of microbes that may have occurred have gone on to great evolutionary success. One simple but very successful type consists of eukaryotic cells that have only mitochondria (no chloroplasts) within them. Mitochondria are the energy factories of eukaryotic cells, and are derived from once free-living bacteria. One theory holds that mitochondria are descended from parasitoid purple bacteria that lived like the modern bacterium *Bdellovibrio*. *Bdellovibrio* attacks its host bacterium the way a virus does, entering the cell and rapidly multiplying until the host cell ruptures, releasing numerous progeny of the invader. (Recall that a parasitoid is like a parasite, only that it almost always kills the host as part of its life cycle.)

In at least one host lineage, the parasitoid bacteria intruder was "tamed" and became, eventually, a rapidly respiring organelle. Its descendants today are called mitochondria; they are the energy factories of the cells they inhabit. This organelle greatly increased the host's ability to metabolize quantities of food. This new cellular development is somewhat equivalent to the transition from cold-bloodedness to warm-bloodedness in vertebrates, and implies a quickening of the metabolism of the components of the newly formed host-organelle combination.

Later, an early protoctist with mitochondria must have ingested a photosynthesizing cyanobacterium. Instead of being digested, however, the cyanobacterium became a welcome guest. The newcomer could diffuse surplus photosynthetically derived food products into the host's cytoplasm. This exchange led to the first photoautotrophic eukaryotes—that is, eukaryotes that could create their own food by using energy from the sun to build sugars out of mere carbon dioxide and water. Prior to this event, all eukaryotes were heterotrophs; they had to find sugars and other organic molecules in their environments from sources such as living bacteria.

The enclosed cyanobacteria evolved over generations into chloroplasts. Chloroplasts may have developed via symbiogenesis as many as six or seven times in five different types of eukaryotic algae,[3] and not all chloroplasts are descended from cyanobacteria. The dinoflagellate *Amphidinium wigrense* has chloroplasts derived from a eukaryotic cryptomonad (a member of the Cryptophyceae, a small group of protoctists). This dinoflagellate thus provides an example of a eukaryote that is endosymbiotic within another eukaryote. The smaller eukaryote forms an organelle within the larger eukaryote.

Even before the origin of mitochondria (and therefore well before the origin of chloroplasts), eukaryotes somehow acquired the motility organelle called the undulipodium. This rapidly beating or undulating organelle (also called the eukaryotic flagellum) comes in two lengths: long (normal undulipodium as in a sperm tail) and short (the bristle-like cilia). The similarity of undulipodia to free-living spirochete bacteria that swim like corkscrews may indicate that undulipodia are derived, via symbiogenesis, from free-living spirochetes.[4] There is no definitive evidence that the undulipodium was acquired by the eukaryotic cell before the other organelles existed, but in any case, acquisition of an undulipodium by a eukaryotic cell made it into the first complex cell able to swim.

Kingdom Taxonomy

According to Lynn Margulis and Karlene V. Schwartz in their book *Five Kingdoms*, the five kingdoms of life are the bacteria, protoctists, animals, fungi, and plants. Ever since R. H. Whittaker's[5] ground-breaking 1959 paper "On the Broad Classification of Organisms," biologists have come to recognize that the old system of dividing life into two kingdoms—plant and animal—is deceptive. A more accurate picture of the diversity of life at the level of kingdom may be obtained by using five kingdoms. These five kingdoms can be defined more precisely on the basis of what types of symbionts they have, once had, or never had.

Bacteria are prokaryotes and, according to proponents of the five kingdom scheme, they are all grouped into one kingdom: Monera. All prokaryotes are found in this one kingdom. Bacteria apparently belong to lineages that never possessed (or perhaps in some cases, once possessed and later lost) endosymbionts.

Monerans can be divided into two or three major groups, but there is controversy about exactly how to divide them up and whether these groups are basic enough to qualify as kingdoms in themselves. Carl Woese and colleagues divide monerans into two kingdoms: true bacteria (eubacteria) and the archaea (formerly archaebacteria). The archaea group includes methanogenic (methane-excreting) bacteria, halobacteria (salt-loving), and some thermoacidophilic bacteria (suited for hot acid springs). They are likely the most ancient lineage of bacteria, well-suited to conditions of a hothouse, anaerobic Earth. Despite their ancient family tree, archaea are viewed by Woese as being more closely related to the original eukaryotic host cell than to the other bacteria. Most organelles, on the other hand, are descended from the eubacteria.

James Lake divides bacteria into three major groups, which he argues should be treated as kingdoms: (1) the eubacteria, (2) a second group that includes halobacteria and methanogens, and (3) eocytes (including some but not all high-temperature, sulfur-metabolizing bacteria). The differences of opinion on the broad outlines of bacterial classification are due mainly to controversies regarding molecular rates of evolution in these bacteria and to the genetic implications of symbiogenesis. Although it is not much discussed, some and perhaps many organisms classified as bacteria may be descended from organisms that were once carriers of symbionts, or alternatively endosymbionts themselves. Symbiogenesis is quite possibly reversible if the symbionts have not yet become obligate and are still able to revert to the independently living state, or if for some reason all of the symbionts die while the host lives on.

The thermophilic bacteria are probably representative of the earliest life both in the sea and, as noted earlier, on land. The two-group and the three-group bacteriologists all agree that the archaebacteria or eocyte bacteria are more closely related to eukaryotes than they are to other bacteria. Archaebacteria have many traits that make them plausible ancestors of the original eukaryotic host cell. For example, archaebacteria have histone proteins similar to those found in eukaryotes. The histones protect DNA from being destroyed by the hot acid found in the hot-spring environments favored by thermophilic bacteria.

Like the present day prokaryote *Thermoplasma,* the original soon-to-be-eukaryotic host cell was probably adapted to less than the present 20 percent concentration of oxygen in the air. After the oxygen crisis two billion years ago these thermophilic bacteria or archaebacteria might have been among the first to develop oxygen-detoxifying enzymes such as superoxide dismutase, because they already had the ability to synthesize the histone enzymes that protected them from hot acid. From the perspective of a single cell, protecting oneself from hot acid and protecting oneself from free oxygen are biochemically related challenges.

Bacteria of various flavors are important to the ancestry of Hypersea (as they are to the ancestry of all complex forms of life) in that they gave rise to the original eukaryotic cell and its various symbionts. Protoctists are also important to Hypersea in that they have always accounted for a major part of the symbionts and parasites internal to the animals, plants, and fungi that give Hypersea its organic structure.

Protoctists are a mixed-bag group of eukaryotes, all of which are single cells or loose communities of single cells. Protoctists include organisms with undulipodia and no mitochondria (such as *Pelomyxa*), those with undulipodia and mitochondria (*Paramecium*), and those forms with undulipodia, mitochondria, and chloroplasts (*Euglena, Chlamydomonas*). The latter group used to be considered either animal-like plants or plant-like animals, but now we simply refer to them as protoctists—they are like themselves and nothing else.

Also included in the protoctists are atypical land organisms such as slime molds. Slime molds are unusual in that, at a particular phase in their life cycle, many types of slime molds occur as a swarm of nuclei within a cytoplasmic mass that coheres and even moves together as a kind of "slug." Other organisms lumped into the kingdom Protoctista are multicellular in the conventional sense of the word. They are the marine algae that include the familiar and delicate sea lettuce *Ulva* and the astonishing sea "forests" of kelp.

Fungus-like protoctists are defined by the fact that they develop undulipodia at some point in their life cycle. Oomycetes or water molds, the microbial villains of the Irish potato famine, are members of the Protoctista. These organisms look very much like true fungi, except for the fact that they have an undulipodiated stage in their life cycle (the unicellular zoospore stage).

An undulipodium at least at some point in life distinguishes so-called water molds from true fungi. True fungi are generally assigned to their own kingdom because they lack any undulipodiated stage in their life cycle, although they may have evolved from a once-undulipodiated group.

Animals have both undulipodia and mitochondria, but lack chloroplasts (unless they acquire them by theft, as in the case of the Hawaiian sea slug *Placobranchus ocellatus,* which photosynthesizes using chloroplasts sucked from the tissue of a marine alga). Also, animals form zygotes from male and female sex cells. The zygote develops to form a blastula (a hollow ball of cells) and later undergoes gastrulation (the hollow ball of cells punches in on one side).

From the point of view of a symbiogeneticist, plants and some protoctists—not humans—are the most "advanced" of organisms, for they contain undulipodia, mitochondria, *and* chloroplasts. Vascular plants, descended from green algae, are the grand success story of eukaryotes, for they create most of the biomass on Earth.

Before leaving discussion of the five kingdoms, we must mention viruses because they will be important to the story of

Hypersea later on. Viruses are nothing more than DNA or RNA, usually packaged in a protein coat. As they are incredibly simple in comparison with even the smallest bacterium, many scientists do not consider viruses to be alive. Indeed, viruses can be crystallized just like many other nonliving organic molecules.

Nevertheless, viruses have many lifelike properties. They are able to self-replicate, although only within the body of appropriate hosts. A virus within its host is considered to be "live" if it is capable of undergoing self-replication, which is usually the case. Interestingly, the closest relative to any particular kind of virus is often not a different type of virus, but rather the DNA of the host itself. It is thought that many if not most viruses are rogue fragments of host DNA that somehow acquired the protein coat that enabled them to disperse and infect other hosts of the same or of a related host species.

With this brief taxonomy now complete, we move on to the situation going on at sea before Hypersea.

The Garden of Ediacara

Our earlier book, *The Emergence of Animals: The Cambrian Breakthrough* (Columbia University Press, 1990), offers an explanation for the geologically sudden appearance of skeletonized animals in the early Paleozoic. The evolutionary puzzle of the origin of animals in general, and skeletonized animals in particular, has a direct bearing on our attempt to understand the colonization of land by complex multicellular life.

The problem begins with this basic question: *When* in Earth history did complex multicellular life actually appear? As we hope to show, the answer fundamentally influences one's view of the history of life and even the broader history of change on the planetary surface. There are two chief competing answers to the basic question.

The first we call the "latecomer" hypothesis. According to

this hypothesis, multicellular life did not appear on Earth until very shortly before the Cambrian explosion, when multicellular organisms suddenly became common and extraordinarily diverse. The Cambrian explosion was, from this perspective, the manifestation of the first successful complex multicellular organisms. They had just evolved in the late Proterozoic, not too much before the beginning of the Cambrian. This view often links increases in evolutionary complexity to stepwise increases in the oxygen content of the atmosphere before (and sometimes even after) the Cambrian.

The second perspective is the "late bloomer" hypothesis. In this view, animal multicellularity was present tens or hundreds of millions of years before the Cambrian boundary, but was not an important biological (and ultimately geological) force until the Cambrian. The reason for the delay was primarily ecological; lengthy heterotrophic food chains involving large organisms did not evolve until the end of the autotroph-dominated Proterozoic. Increases in atmospheric oxygen and drops in temperature may have encouraged the initial development of eukaryotic cells, but the evolution of complex multicellularity and the emergence of animals was *not* directly linked to the attainment of threshold levels of atmospheric oxygen, temperature of the atmosphere, and so on.

The simplicity of the latecomer hypothesis, in which complex life does not appear until the Cambrian itself, is attractive. Under this hypothesis no large multicellular animals occur before a certain key and discrete evolutionary event, perhaps triggered by some change in the atmosphere. This physical or chemical change is quickly followed by the evolution of not just a few types of emerging animals, but by a swarm of diversification—the "Cambrian explosion."

Alas, the latecomer hypothesis has apparently fatal difficulties. For the latecomer hypothesis to be correct, all of the fossils before 600 million years ago (approximately the time of the origin of the Ediacaran biota) should be fossils of bacteria and protoctists, and none should be multicellular animals or plants.

Our current understanding of the fossil record of Ediacaran time and earlier would seem to eviscerate the latecomer hypothesis. Fossils of probable multicellular eukaryotes are scattered throughout marine rocks up to 1.3 billion years in age.[6] The latecomer hypothesis appears frail indeed in the light of such old but complex fossils, especially when most paleontologists agree that probable oxygen-producing multicellular eukaryotes themselves occur as fossils (acritarchs) nearly two billion years old.

Proponents of the latecomer hypothesis have argued that new cell types can be added only slowly to the developmental programs of multicellular organisms. Algae have only several cell types, but a typical invertebrate has about fifty different types of cells, and vertebrates (including humans) have several hundred types. If it takes a long time to evolve a new cell type and to integrate it into the genomic and developmental program of the multicellular organism, then long stretches of pre-Cambrian time might well be necessary to attain the fifty or so cell types required for an animal. This could explain the long delay between the origin of eukaryotes and the emergence of animals. Complex multicellularity would not appear on the scene until some threshold number of cell types had been passed, a threshold that must have happened right before the Cambrian.

The cell type theory (a version of the latecomer hypothesis) fails, however, on the following grounds: Why would a billion and a half years be required to go from, say, *Eosphaera* to members of the Ediacaran biota? Fossil evidence suggests that Ediacaran organisms (figure 3.2) could not have had too many more cell types than *Eosphaera* had, thus cramming the appearance of the majority of animal cell types to the tail end of the Proterozoic (latest pre-Cambrian) and voiding the advantages of the cell type theory. Of course, complex multicellularity could have evolved more than once, only to be snuffed out during pre-Cambrian mass extinction events. But this argument requires one to believe either that no fossils were preserved or that they are yet undiscovered. Neither prospect is likely considering that the fossil record shows a long, uninterrupted span of existence of certain probable eukaryotes, like the acritarchs and the carbonaceous megafossils.

A version of the late bloomer hypothesis, rather, is the one we supported in *The Emergence of Animals*. In that book the late bloomer hypothesis is called the Garden of Ediacara hypothesis, in reference to the Ediacaran biota, a group of pre-Cambrian soft-bodied animal-like organisms which appear to use self-sufficient feeding strategies such as photoautotrophy and chemoautotrophy. Self-sufficient feeding strategies meant that nobody had to eat anybody else. Virtually all the multicellular organisms coexisted peacefully.

The Garden of Ediacara was the first time that animals or

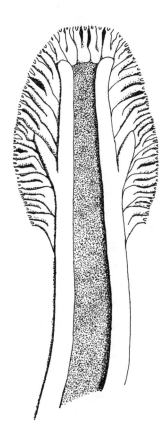

FIGURE 3.2

Enigmatic Ediacaran organism. The fossil is from 600-million-year-old strata in southern Australia. Note bifurcating branches of radial canals. (Length of specimen is 8.1 cm.) Members of the Ediacaran biota were the first large multicellular organisms and appeared at least 10 million years before the Cambrian explosion.

animal-like organisms attained large size. We believe they were assisted in their growth by photosymbionts and chemosymbionts—indeed, symbiosis was the key that unlocked this first significant expression of animal or animal-like multicellularity.

Gigantism in modern marine animals is often associated with the presence of photosymbionts and chemosymbionts. Paleontologist Bruce Runnegar argues that Ediacaran tubular fossils from Australia, along with the leaflike Ediacaran frond fossil *Phyllozoan*, resemble the giant, chemosymbiotic tube worms of modern deep sea vents, and that these fossils support the Garden of Ediacara hypothesis because the similarities are not merely superficial. The massive Ediacaran fossil from Namibia called *Ernietta*, shaped something like a ten-centimeter-thick baby elephant's foot, is often found with large pyrite (iron sulfide) crystals at its base, further supporting the possibility that Ediacaran organisms used chemosymbiosis (i.e., sulfate-reducing bacterial symbionts) as a mode of nutrition, since pyrite is often associated with fossil deposits of organisms known to have lived in this way.

Fall of the Garden of Ediacara

The late bloomer or Garden of Ediacara hypothesis attributes the delay in the appearance of metazoans to the existence of a photosynthetic or chemosynthetic autotroph-dominated sea floor during the pre-Cambrian. This self-sufficient global marine ecosystem was literally torn to shreds by the advent of large burrowers and predators. By their activities, these new, predatory forms created lengthy trophic chains, bringing an end to the Garden of Ediacara and thereby remaking the biotic face of the earth.

The marine world existed for tens of millions of years with large animals or animal-like organisms (the Ediacaran biota) but without megascopic predators, not because of an oxygen deficit nor because complex multicellularity had not yet evolved, but because predators with eyes and the ability to seek out and attack mobile prey had not evolved. Large predators had not yet been invented. When they did appear in or immediately before the Cambrian, the frantic response by the prey organisms at risk unlocked a vast, previously untapped reservoir of evolutionary potential, and the skeletonized animal phyla of the Cambrian followed in short order.

A causal link may exist between the Cambrian evolutionary explosion and marine nutrient changes brought about by plate tectonic rearrangements. Geological evidence indicates that about 700 million years ago there existed a supercontinent Rodinia, composed of all or nearly all of the present-day continents.[7] Rodinia was torn to bits by continental drift not long before the Cambrian, and in a second phase of drifting some of the orphaned pieces collided to form a smaller supercontinent called Gondwana (composed of the present-day southern continents). The first phase of continental drift seems to coincide with a time when essential nutrients such as phosphorus and nitrogen were low in marine waters but dissolved levels of carbohydrates were high—this would be the Garden of Ediacara, the Age of Autotrophy.[8] The second phase of continental drift and continental collision seems to be synchronous with a rise in global sea level (also called a transgression) and, since collision-associated andesitic volcanoes form nutrient-rich sediments when they erode, also to be synchronous with the injection of abundant phosphorus and other mineral nutrients into the marine realm. This nutrient pulse is a likely trigger for the emergence of large predatory animals and the ensuing large-scale disturbance of the marine ecosystem.

We see this nutrient pulse as a much more plausible trigger for the emergence of skeletonized animals than, say, the crossing of a threshold oxygen level in the atmosphere or some other purely physical change in the environment. Simon Conway Morris[9] suggests that the apparently rubbery cuticle of Ediacaran organisms was an adaptation to protect their tissues from very high levels of atmospheric and hydrospheric oxygen—so the tide of scientific opinion seems to be turning against the oxygen threshold hypothesis. Indeed, animals can and have evolved to thrive under a variety of oxygen levels, but it is well known that improved access to needed nutrients such as phosphorus and nitrogen can substantially alter the course of evolutionary change. (We will supply what we regard as convincing evidence of this effect on land when we discuss hypermarine upwelling in chapter 9.)

Large size may have protected the Ediacaran biota from the tiny predators of the day; the Garden of Ediacara interval may have actually been a lag while the predators caught up in size. When large, animal, brainy predators evolved that were capable of seeking out and capturing mobile prey, all others were suddenly at risk and the selective pressure for skeletons and other forms of protection became intense. Thus followed the "Cambrian explosion" of skeletonized metazoa, at a time of which it truly can be said, in the words of William Stolzenburg,[10] that "life got hard."

Only a single animal type with a hard skeleton evidently used as antipredatory defense shows up in the fossil record before the Cambrian (specifically, during the Vendian period). This is the oldest shelly fossil *Cloudina*, a globally widespread form. *Cloudina* has a multilayered shell structure that looks a bit like a tall stack of disposable drinking cups. Cloudinids have been found with possible bore holes, suggesting predatory attack. Fossils with possible bore holes are, however, exceedingly rare, occurring in only one known locality. *Cloudina*, on the other hand, is very abundant, suggesting that the Vendian predators of *Cloudina*, if they existed, were the exception rather than the rule. In any case, these predators attacked a stationary prey, and no evidence indicates that they were sophisticated enough to chase prey around the sea floor.

Before proceeding further, it would be best to review the three traditional arguments that have been presented linking predation pressure and related ecological interaction to the Cambrian appearance of skeletonized animals. The first and oldest of the arguments or "factors" is that Cambrian skeletons appeared as a simple defense against predators. No big increases in diversity (number of different types of organisms) or disparity (taxonomic distance between unrelated animals) is necessarily predicted in this case. Proponents of this theory have argued that no real diversity spurt occurred at all: the Cambrian came in with a "whimper not a bang." Although it seems to imply an unrealistically long pre-Cambrian history for animal phyla, the idea is probably at least partially correct, even if it cannot by itself explain the diversity of skeleton types in the Cambrian.

This brings us to the second argument. Perhaps the apparent increase in diversity was real, having been caused by cropping pressure—that is, attrition of a variety of herbivores by predators. Steven M. Stanley is a proponent of the cropping pressure school of thought. His intuitive leap was to compare modern ecological studies off the West Coast of North America with the Cambrian diversification. Stanley contended that the cropping pressure of predators and grazers not only maintained diversity (as had been demonstrated for a starfish-dominated community from rocky shores) but that it triggered the evolution of diverse forms. Cambrian predators generated the diversity by clearing out enough ecological space to encourage evolution.

Stanley's idea has been criticized on four points raised by Richard Cowan.[11] First, Cowan notes the probability microbial predation existed a half billion years before the Cambrian. Second, Cowan contends that the large Cambrian predators were rare. (This objection is now dated, as we will show.) Third, he cites a lack of evidence that cropping pressure can *generate* diversity. And fourth, he harbors a suspicion that the rocky coast studied in the starfish research may have been a special case since diversity can be lowered or raised depending on the identity of the top carnivore of the community in question.

The first objection to the cropping pressure idea is not, in our view, valid because it ignores possible delays in the advent of predation at the megascopic scale. As suggested earlier, there may have been a time lag (lasting millions or even billions of years) while predators caught up to this size. In other words, herbivores and autotrophs can evolve to such a size that they place themselves out of the size range were they would be endangered by attacks from smaller, contemporary predators. The predators could then be put in a position of having to play evolutionary catch-up. Microbial predators adapted for preying on other microbes do not necessarily affect centimeter-sized organisms. After all, one possible explanation for the comparatively gigantic sizes of some of the Ediacaran biota specimens would be that these creatures used large body size as a strategy to escape microscopic predators.

Cowan's second objection is of less concern now that the meter-long Cambrian predator *Anomalocaris*, formerly known only from the northern hemisphere, has been found in Cambrian rocks of Australia; it is thus a worldwide predator. This creature had a hideous pineapple-ring-shaped mouth lined with teeth, and was without question the *Tyrannosaurus rex* of the Cambrian sea floor.

The third criticism is valid; cropping pressure probably does not open up *enough* ecological niche space to allow evolutionary *generation* of novelty, although it is fair to suggest the possibility that predation pressure could maintain diversity. We believe that this same cropping pressure eventually drove shallow-water algae landward to explore and eventually colonize land environments.

Cowan's fourth point is also valid; introduction of new predators probably leads to diversity drops more often than to diversity increases. The effect of a predator on diversity can depend heavily on whether or not the prey forms at risk are driven to extinction by predation.

The third factor is the "transcendent function switch." This factor involves a rapid change in function of an antipredatory adaptation that leads to new lifestyles that might never have been attempted if predatory pressure had not been applied. These function changes occur almost simultaneously in unrelated groups of organisms. For example, the evolution of hard, protective, calcitic plates in a sessile, formerly soft-bodied, filter-feeding animal can lead to the evolution of a stem composed of platelets, thus raising the filter-feeder up into the water column where there is access to more food or light. Crinoids, or "sea lilies," are an example of this evolutionary development. Thus, prey response to predatory pressure transcends the biotic "arms race" between predators and their prey, and leads to major innovations in the preys' lineages. Clearly, the transcendent function switch requires some component of simple predator defense to initiate the evolutionary process.

The ternary diagram of figure 3.3 shows the three factors discussed above. Stanley's hypothesis would fall near the cropping corner; the Whimper hypothesis in the simple predator defense corner, and the Garden of Ediacara hypothesis as shown close to the function switch corner. (The shaded rim indicates an area in which no valid hypothesis could lie.) Like function switch, cropping pressure cannot work without at least a small component of simple predator defense; that is why only the whimper hypothesis appears exactly on a corner. A middle-of-the-road position can, however, be found somewhere near the midpoint of the ternary diagram (center of the triangle). The hypothesis in that position would be that the Cambrian

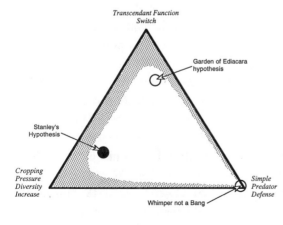

FIGURE 3.3

Cambrian Hypotheses Ternary Diagram. The stippled areas block off zones of impermissible hypotheses.

explosion was attributable equally to the combined influence of predator defense, cropping pressure, and function switch.

Figure 3.3 is useful as a concept plotting diagram, but is there any way to really test or assess the relative importance of the three factors? This is an important topic for future research, and we believe that in principle it can be done. But we are reviewing the puzzle of the Cambrian explosion only as an important prelude to understanding the emergence of Hypersea.

Moreover, the similarities among the three posited factors are more important than their disparities: all assume that ecological interactions played important roles in the Cambrian explosion. This is a major assumption, but justified, because the marine biosphere would not have been likely to omit heterotrophic multicellularity for a half billion or more years (the gap between the origin of eukaryotes and the emergence of large predatory animals), unless there were a fundamental reason for heterotrophic multicellularity to come when it did. We believe that explanation has to do with the appearance of aggressive heterotrophy at the scale of centimeter-sized organisms, and with what must have been the millions-of-years-old stability of a marine biosphere populated with large and small autotrophic organisms, from *Eosphaera* to the Ediacaran biota. The heterotrophic system finally prevailed, but even today marine and aquatic animals are strongly drawn to symbiotic-autotrophic lifestyles. "Nature red in tooth and claw," is clearly an incomplete description of ecosystems possible and present on Earth.

The idea that predators were evolutionarily important during the Cambrian picked up significant support recently, with the discovery of early Cambrian foraminifera from West Africa. Foraminifera are amoeba-like protoctists that secrete carbonate coverings resembling fanciful, miniature snail shells. Foraminiferan fossils so early in the fossil record are important evidence that organisms not closely related to animals also developed hard parts to inhibit predators at or near the Cambrian boundary.[12] This makes the Cambrian explosion of protective coverings a multi-kingdom affair. Calcium ions had been abundantly available for billions of years in seawater, but only at this time in Earth history did marine organisms (animals and protoctists) begin to use them for skeletons.

Marine multicellular algae (another group of protoctists) must have felt the same pressure of predation, or more specifically, grazing, that was felt by the early prey animals. One way for a marine alga to avoid being eaten is for it to develop a calcareous skeleton. Indeed, calcareous algae first appeared in the late Proterozoic and became very abundant in the Cambrian.

Another way for marine algae to avoid being chewed on is to migrate to an environment where the herbivores do not exist. In the early Paleozoic the land was virgin territory, save for microbial crusts and mats. But getting there required some unusual physiological changes.

FOUR

Land Life Before Hypersea

The origins of terrestrial life are veiled in mystery. The nature of the mystery boils down to two words: insufficient data. It is a situation not dissimilar to the inadequacy of data in the last century pertaining to marine ancestors of the Cambrian faunas, before the discovery of the Ediacaran forms.

The First Fossils

Direct evidence in the form of microbial fossils of land life from rocks up to 1.2 billion years old has been reported by L. Paul Knauth and Robert J. Horodyski.[1] They discovered Proterozoic land microfossils from several buried terra rossas from the southwestern United States. A terra rossa is a residual red clay deposit mantling limestone bedrock; the term is Italian for "red earth." Terra rossas form when limestone bedrock is exposed to long-term weathering. Exposure to air brings oxidation, dissolution, and erosion of limestone, and leaves a residuum of clay. Soils may even develop.

The red color of the terra rossa is due to its content of oxidized iron. Knauth and Horodyski's Proterozoic terra rossas were buried and later exposed by mountain uplift. The fossils in the terra rossas occur in secondary silica deposits where chert nodules formed in voids. Some of the fossil nodules were even eroded out of the developing terra rossa during the Proterozoic, and shortly afterward were sculpted by wind and particle abrasion in a desert environment. Such sculpted nodules (called ventifacts) are unusual geological specimens, usually hand-sized or smaller with smoothed, polished-feeling crests and ridges. Some of the Proterozoic land microfossils from the southwestern United States occur in ventifacts, and the rest occur in chert still contained at the weathering surface or within the terra rossa of the ancient, exposed limestone surface.

We accept Knauth and Horodyski's fossils as authentic relics of early land life. The fossils are simple filaments resembling modern cyanobacteria. Some of the specimens are clustered together to form a feltlike mat of filaments, suggesting to Knauth and Horodyski that the fossils represent *in situ* specimens and not microbes transported in from another habitat. Indeed, the iron oxides of the terra rossa at the top of the Mescal Limestone (one of the two terra rossa deposits studied by Knauth and Horodyski) may have been formed by iron-oxidizing bacteria.

In spite of the great age (1.2 billion years) of Knauth and Horodyski's finds, there is reason to believe that organisms colonized land even earlier. Hardy marine bacteria were almost certainly able to colonize land several billion years ago. Today the descendants of these pioneering organisms are living in some of the harshest terrestrial environments one can imagine (these will be discussed shortly). Theoretical considerations imply that there had to be significant land life during the pre-Cambrian. As noted in the previous chapter, land life appears to have existed since very early in Earth history, based on the intensity of pre-Cambrian weathering. Biotic enhancement of weathering played a significant role in improving the habitability of the early Earth by bringing down global temperatures.

The most attractive candidates for the earliest land colonizers are the thermophilic bacteria, or thermophiles. These bacteria are autotrophs, today primarily found in hot springs. They build sugars from carbon dioxide and the hydrogen that they extract from the hydrogen sulfide that is plentiful in hot springs. In all likelihood, land thermophiles evolved very early, possibly as far back as the origin of life itself. They rapidly colonized the land and soon brought on biotic enhancement of weathering and associated environmental changes such as soil stabilization. This resulted, according to David Schwartzman and Tyler Volk, in a rapid drop in global temperature soon after the origin of life, and the creation of a new steady-state balance in the amount of carbon held in Earth's major carbon reservoirs: the atmosphere, crust, and ocean.

As noted in chapter 2, the early thermophilic microbes largely put themselves "out of business" by cooling the Earth's surface below 20°C. They still exist, of course; we admire them on visits to Yellowstone's thermal pools, but we just don't come upon them in the course of our daily lives. Other life forms eventually evolved at lower temperatures, but even at lower global averages thermophiles might have been the only organisms that could thrive in exposed areas of land at certain times of the day. The terrestrial environment is indeed much harsher than the sea.

This is not to say, however, that more complex life could not in principle have evolved at the high early temperatures. Perhaps complexity simply did not have enough time to evolve before the global climate had substantially cooled. In a curious reversal, avian and mammalian warm-bloodedness brings our cells back to early Earth surface temperatures.

Modern Analogs of Early Land Ecologies

Weathering, or physical and chemical breakdown of rock, is a soil-building process. Since the mere presence of soil (as opposed to bare rock) enhances chemical weathering by providing an increased surface area of rock fragments as soil particles, land-dwelling microbes would have, by their stabilizing effect, greatly increased both the pre-Cambrian weathering rate and the accumulation rate of early soil.

Today, vegetation from microbial crusts to higher plants

plays a crucial role in building soils and slowing soil erosion. Desert regions have cryptogamic soils, stabilized by a community of bacteria and lichens that flourishes when it rains and then dries into a wind-resistant (but footfall-fragile) crust. Managers of the great desert parks of Arches and Canyonlands in Utah post signs warning visitors to stay on the paths in order to protect cryptogamic crusts from trampling. Soil ecologies like this, though peculiar today, are plausible models for pre-Cambrian soils. Their bacterial, algal, and fungal components, of course, make them taxonomically different from the kinds of strictly bacterial crusts that might have been the earliest colonizers. But they seem eerily adapted for an earlier, less competitive era: cryptogamic crusts are found mostly in deserts today not because that is their preference, but because deserts are the only places where they are not crowded out.

A more taxonomically exact analog for pre-Cambrian land life is also found primarily in deserts, on surfaces too hostile even for crustose lichens. "Rock varnish" owes to the workings of bacteria. Black or rusty streaks and stains on seemingly bare and shiny sandstones, rock varnish is actually a thin coating of clay cemented to rocks by manganese and iron oxides. In the early 1980s scientists confirmed that rock varnish was formed by bacteria that concentrate mineral oxides in an opaque film. The bacterial genera *Metallogenium* and *Arthrobacter* are specific examples.[2] Research in the coastal desert of Peru has shown that dew on rocks can release manganese, which is then precipitated by the bacteria to form a black varnish. High alkalinity inhibits the manganese concentrators, and a reddish varnish with a low manganese-to-iron ratio results. The rust-colored varnish is an iron oxide clay dust that is less well secured to the rock surface than is the manganese-rich varnish.

Varnish-forming bacteria are certainly candidates for pre-Cambrian land life, particularly considering the abundance of bare rock surfaces at the time. A marine form (*Eoastrion simplex*) morphologically identical to *Metallogenium*, is known from the two-billion-year-old Gunflint Iron Formation. *Eoas-*

trion ("dawn star") looks, as its name suggests, like a microscopic starburst formed of metal oxides.

The Ecology and Fossil Record of Fungi

"Toadstool Grows, and Grows" read the Associated Press headline in the April 2, 1992 issue of our local newspaper; "There's a Horror Movie in Here Someplace," announced a national tabloid. It is not often that a fungus makes it into the national news, but this was an exceptional fungus. Using genetic analysis, scientists had discovered a single 1,500-year-old living fungus whose underground mass of threadlike filaments (called hyphae) spread through fifteen hectares of soil in Iron City, Michigan. This specimen of *Armillaria bulbosa*, the honey mushroom, was claimed to be one of the oldest and largest organisms on earth, covering a roughly triangular area bigger than five football fields. The species is a saprotroph on dead or diseased wood, and the fungal mass probably weighs at least a hundred tons, or about the same mass as an adult blue whale.[3] An even larger specimen of the related species *Armillaria ostoyae* (a parasite of conifers) has been recently reported to inhabit 6.25 square kilometers in Washington state. This fungus is killing the Douglas Fir trees in the area, and foresters are in dismay, not knowing how to contain it. The fungus forms large mats of hyphae beneath the bark of living trees, consuming the living tissue and destroying the tree in the process.

The discovery of the fungal giants is a vindication of sorts for mycologists. Tom Bruns, a plant pathologist at the University of California at Berkeley, feels that fungi such as these have been neglected and that "some of the major players in terrestrial ecosystems have been ignored by the scientific community." No one is now suggesting that the Michigan and Washington cases are unique. Honey fungi (genus *Armillaria*) are a widespread and serious disease of garden shrubs and trees.

The search for giant fungi has only just begun, and we suspect that such huge individuals may be the rule rather than the exception.

The term *fungus* can be used with two different meanings. The more strict denotation of fungus is a term for a kingdom level taxonomic group—the kingdom Fungi. As defined by Margulis, the fungi in this sense are a group of eukaryotes descended from an ancestral aerobic eukaryote ancestor with mitochondria and undulipodia. Unlike the plant and animal kingdoms, the fungi have secondarily lost the ability to form undulipodia, the whiplike appendages so helpful for motility in aqueous environments.

The second and more common usage of the term fungus is in an ecological context: a fungus is a type of eukaryotic (and occasionally prokaryotic) organism that forms a mesh of hyphae or cytoplasm-filled filaments and feeds by absorbing nutrients from other organisms or the remains of organisms. The hyphae may form a downy mass called a mycelium. True fungi are of course included in this definition, but so are the oomycetes or water molds, which are actually members of the kingdom Protoctista. (Water molds form undulipodiated unicells during part of their life cycle.)

Oomycetes (water molds) and true fungi have undergone, with respect to one another, a tremendous amount of convergent evolution. These groups are not at all closely related, and yet they share a nearly identical mycelial form. Even their reproductive organs appear similar.

Since oomycetes are capable of forming undulipodia, they do not belong in the kingdom Fungi. They can, however, be referred to as fungi in the second or ecological sense. This same divergence in usage can be seen in the word *plant*, which taxonomically refers to members of the kingdom Plantae (mostly terrestrial and always multicellular organisms with chloroplasts, mitochondria, and undulipodia), but which in decades past was often used in an ecological context to refer to any moneran, protoctist, fungus, or green plant—that is, any organism that is not an animal! For instance, the Associated Press article reporting the giant honey mushroom referred to the organism as a "plant." Needless to say, we discourage this latter usage.

The published fossil record of fungi is extraordinarily sparse compared to other, more popular groups of fossil organisms. Nevertheless, well over 500 species of fossil fungi (in the ecological sense) have been described. Occurrences of fossil fungi are generally restricted to coals, infected animals in amber, or to sites of exceptional soft-bodied preservation. The oldest known fossils of true fungi are from Upper Silurian sediments in the Ludlovian of Sweden.[4] The Devonian Rhynie Chert of Scotland (mentioned in chapter 1) contains a diverse collection of fossil fungi. Later still, Pennsylvanian fungi of the ascomycete type have been identified in both coal balls and stratified coal in North America. The ascomycetes were recognized by their flask-shaped, nonopening fruiting bodies (called cleistothecia), which in turn contain saclike cells that hold the spores.

Coprinus is a common modern mushroom whose name literally means "feces eater," in reference to one of its preferred habitats. The fossil species *Coprinites dominicana*, known from Cenozoic amber from the Dominican Republic (early Oligocene to late Eocene), is the earliest known fossil of a mushroom. To the nonspecialist it is virtually identical to a modern mushroom. No doubt earlier mushrooms existed, perhaps as far back as 200 million years ago, but none older are yet known as fossils.

A Brief Taxonomy of Fungi

In the lichen part of this chapter and in later chapters that deal with the emergence and development of Hypersea, it will be useful to understand the distinctions in fungal forms and physiologies and to be familiar with basic terminology. There

are three main types of true fungi: ascomycetes, basidiomycetes, zygomycetes, and deuteromycetes.

In the ascomycetes, spores are formed in and enclosed by a
saclike cell called an ascus (plural, asci). Mildew, truffles,
baking yeast, and morels are familiar ascomycete fungi. The
Pennsylvanian coal ball fungus is a fossil ascomycete. Ascomycetes can form spores asexually or sexually. Sexually produced
spores are called ascospores. Ascomycete fungi might be accused of "overpackaging" these ascospores—clusters of ascospores are held together in an elongate, saclike structure called
an ascus. In turn, clusters of asci are enclosed in fruiting bodies
called ascocarps. The cleistothecium of the Pennsylvanian coal
ball fungus is a type of ascocarp.

Basidiomycetes are the most familiar fungi. Their above-
ground reproductive structures, mushrooms, are variously consumed as delicacies, sought for hallucinogenic properties, or
feared as poison. They are the fungi of fairy rings and the
toadstool that Alice tasted in Wonderland. Basidiomycetes include the grocery store mushroom (*Agaricus bisporus*), *Coprinus*
inky cap mushrooms, *Amanita* (the "Angel of Death"), and
Boletus (some forms are edible), as well as the common shelf
fungus *Polyporus* (figure 4.1). All these fungi are distinguished
by their spore-bearing basidia, which look like tiny tuning
forks. Below ground (or hidden in the body of their host), the
mycelia of basidiomycetes contain interesting clamp connections
between partitioned sections; the partitions themselves (septa)
are perforated by a centrally located, barrel-shaped structure
called a dolipore septum. This pore is covered by a membranous pore cap.

Deuteromycetes or "imperfect fungi" are in fact a mixed bag
of fungi that includes the sexless forms of ascomycetes and
basidiomycetes. They still produce spores, but the spores are
produced asexually, and are called conidia (singular, conidium). Some deuteromycetes do not form spores of any kind,
and rely on other asexual modes of reproduction. Many deuteromycetes do, however, undergo hyphal anastomosis—a phe-

nomenon important to Hypersea. In this process hyphae from different strains of fungus (of the same species) fuse to form cytoplasmic connections. Fusion can be followed by a "parasexual" event in which nuclei from the different strains encounter one another, fuse, and thereby establish a new genetic strain.

The most familiar deuteromycete is *Penicillium notatum*, the source of the drug penicillin. Another common deuteromycete is *Alternaria*, easily recognized under the microscope by its conidia, which resemble segmented bowling pins. *Alternaria* can cause leaf spot disease in plants. Deuteromycetes are not the only forms responsible for leaf spot diseases, however; leaf spot can also be formed by ascomycetes and by bacteria.

The date of origin of both ascomycete and basidiomycete fungi is currently being debated. Some scientists believe that both groups arose well before the origin of plants; others (basing their opinion on rather unreliable molecular clocks) have them originating after the origin of land plants. Fossils of the mycorrhizal glomacaean fungi mentioned in chapter 1 suggest that this ancient group is neither an ascomycete nor a

FIGURE 4.1
Polyporus basidiocarps growing on a tree.
Mark McMenamin for scale.

basidiomycete. It is probably related to the common ancestor of both ascomycete and basidiomycete fungi. The fossil record is mute on the origins of these groups. The oldest fossil shelf fungus, *Phellinites*, is from the Jurassic about 165 million years ago.[5] Clamp connections between cells characteristic of basidiomycetes are visible inside the fossil of a Coal Age fern, about 290 million years old.

The Intrepid Lichens

People take symbiosis seriously in lichens, but then they dismiss lichens as unimportant. —*Lynn Margulis, 1992*

A close look at lichens reveals that their true nature is less idyllic than we presumed, but I suppose that might be said about most biological associations. —*Vernon Ahmadjian, 1982*

Some of the most commonly encountered, if not terribly conspicuous, forms of fungi on land are lichens (figure 4.2). Lichens are important colonizers of abiotic environments, such as freshly erupted lavas, and most seem to do best where air is unpolluted. Lichens are the most well-known composite organisms. They consist of a fungal member called the mycobiont, and a photosynthesizing partner called the phycobiont. The lichen species is usually named for the mycobiont, which in most lichens is an ascomycete fungus, but basidiomycete and deuteromycte fungi can also form lichens.

Ascomycetes and lichens both have very poor, nearly nonexistent fossil records. Two species of lichens have been reported in the Eocene and Oligocene of the Baltic region, attached to the leaves of angiosperm plants preserved in amber. Virtually the only known Paleozoic fossil lichen is *Spongiophyton*, a Devonian form we will discuss in more detail later. *Spongiophyton* was originally thought to be a fossil plant.

The phycobiont in a lichen can be either a prokaryotic cyanobacterium ("blue-green alga") or a eukaryotic chlorophyte (a green alga). In some lichens both types of phycobionts are present. The cyanobacterium is thought to have a nitrogen-fixing as well as a photosynthetic role.

A typical lichen is composed of four layers. The outer layer (the upper cortex) faces the sun. It is composed of fungal cells, and it forms a protective outer coat. Immediately below the upper cortex is the algal layer, were the photosynthetic activity takes place. Below this is the medulla, a storage area formed by scattered fungal hyphae. The lowest layer is the lower cortex, which forms structures like root hairs that attach to the substrate. The lichen can form fruiting bodies, and these fruiting bodies release tiny airborne propagules called soredia. Soredia consist of an algal cell wrapped by fungal filaments.

There can be more to a lichen than simply fungi and cyanobacteria or green algae. David Schwartzman has found, and Norrie Robbins has identified, chrysophyte algae adhering to the inner wall of a lichen cortex. Chrysophytes are another kind of photosynthetic, eukaryotic alga distinct from the chlo-

FIGURE 4.2

Lichens growing on the bark of a red maple tree (*Acer rubrum*). Width of view is 85 mm.

rophytes. They are often covered by minuscule siliceous scales. Because they are typically planktonic marine or freshwater organisms, it seems odd that a species of chrysophyte would find a terrestrial form like a lichen as suitable habitat. But when they do, the lichen itself becomes even more composite than usual. It is not known whether or not fungal hyphae penetrate the chrysophyte alga in the way they penetrate the cells of the more usual phycobionts.

A rancorous debate has been going on for most of this century regarding the nature of the lichen symbiosis. On one side of the argument are the symbiogeneticists, who see the biological association of the various "bionts" in a lichen primarily as a sort of "mutual aid society." On the other side are those, beginning in 1902 with the Russian biologist Alexander Elenkin, who view the lichen symbiosis as a case of controlled parasitism. Today some scientists feel that the debate over the dual nature was resolved in favor of the concept of fungal parasitism, but this conclusion is by no means certain.

In our view, a lichen is not just a passive community of mutual "friends," for the fungus appears to control the association—that is, it appears to be able to kill the phycobiont at will. The benefits of the lichen symbiosis appear to be tipped in favor of the fungus. In many species, the mycobiont actually invades the cell of the phycobiont with specialized hyphae called haustoria (absorptive hyphal branches that penetrate host cells). The role of haustoria in the symbiosis is not clear, but it is known that haustoria can kill algal cells that are not healthy.

Haustoria are not observed in all lichens—*Coenogonium interplexum*, for example. The cell walls of its ascomycete mycobiont do, however, appear thinner at points of contact with the phycobiont (in this case the filamentous green alga *Trentepohlia*), and several branched fungal hyphae can on occasion completely ensheath the algal filaments. Although in this case there is no actual penetration of the phycobiont cells by the mycobiont, the association is clearly a lichen because there is some modification of *Trentepohlia* in comparison to its free-living form and because the lichen growth form is structurally distinct from the growth forms of the two symbionts in their free-living states.

Whichever partner in truth captures the greater part of the spoils or controls the association, the phycobionts derive at least one major benefit: airborne, hardy soredia carry the phycobionts to areas where they might not otherwise grow. Lichens have been shown to survive a temperature of minus 198°C. In the Antarctic, active photosynthesis has been recorded in lichens at ambient temperatures of minus 18.5°C. Unquestionably, the lichen is, in a biotic sense, greater than the sum of its component parts.

Perhaps it is more appropriate to think of the fungi as actively "farming" the phycobionts for photosynthetically derived food products, rather than parasitizing the phycobiont outright. To continue the metaphor, phycobionts may from time to time go feral. Some phycobionts probably, on occasion, escape the clutches of the fungus and establish pure algal or pure cyanobacterial colonies. One can imagine a situation where a lichen is subjected to some sort of fungicide or a fungistatic compound (such as those secreted by ants) that damages the mycobiont but leaves the phycobiont unharmed.

Lichenization is not restricted to terrestrial environments. A marine lichen has been reported that has the brown alga *Ascophyllum nodosum* as phycobiont and a secondarily marine ascomycete fungus as mycobiont. Thomas N. Taylor notes that there is an ever-expanding data base that suggests the presence of marine fungal symbioses prior to the "invasion" of land.

Living in the Rock and "Lichen" It

In 1974 biologists E. Imre Friedmann and Roseli Ocampo-Friedmann made an astonishing discovery in one of the most inhospitable environments on Earth: the dry valleys of Antarctica. The dry valleys region is really a polar desert, positioned south of Antarctica's Victoria Land. It consists of 5,000 square kilometers of deep, U-shaped glacier-cut valleys flanked by dolomite and sandstone highlands. Frigid, desiccating blasts of wind from Antarctica's continental ice cap ablate or scour away any ice or snow cover that may (rarely) form or fall. Rock in the dry valley is alternately frozen and thawed, not uncommonly in cycles of under an hour because of the frequent wind gusts. The rock surfaces, warmed by summer sunlight to temperatures above the melting point of ice, plunge below freezing at each gust.

The dry valleys present an apparently lifeless landscape, with absolutely no animals—not even insects—and no plants. But a mere two or three millimeters beneath the barren rock surface is a thriving, tenacious community of fungal filaments and their algal and bacterial associates. These cryptoendolithic (literally, "hidden in the rock") lichens have taken refuge inside the porous Antarctic sandstone, where they are protected from the violent temperature fluctuations of the rock surface and the air surrounding it. The lichens (and colorless bacteria who share the same pores) use the rock, in effect, as a miniature greenhouse, owing to the fact that sufficient light for photosynthesis can pass through the sandstone's translucent grains of quartz. Also, the outer layer of rock consists of a thin silicified crust, which creates a protected, solar-heated interior space. The porous sandstone also traps water. On the very rare occasions when a snowfall melts on the rock surface, it is absorbed into the pores.

Friedmann and Ocampo-Friedmann first noticed these lichens when they cut open a sandstone sample from the dry valley and noticed distinct colored bands below the surface of the rock. These bands represent concentric microbial zones. In aggregate, the zones are only slightly more than a centimeter in thickness. The outermost zone is black and composed mostly of fungi. This zone absorbs heat and light. The second zone is composed of colorless fungal hyphae (filaments). The third and innermost zone consists of a green band of algae. Friedmann compares the community to a rain forest, with the green-zone organisms at the bottom living in deep shadow, having to survive with whatever "light they can get."

On the other hand, the organisms are clearly in association with one another. The intense light of austral summer might be too much for the algal partner without protection by the fungal "parasol." In this lichen there is a slight physical separation between the phycobiont (the green zone) and the mycobiont (the black and colorless zones). But the partners can depend on the stable system of pores in the rock to guide whatever transfer of fluids and nutrients their arm's length intimacy requires.

The Antarctic ecological community in any one place (indeed, within any one rock) is likely hundreds of years old, because the silicified rock crust that encases the lichen community forms by slow geological processes. These communities are candidates for the oldest living lichens. It is not clear, however, which came first—the lichen or the silicified crust. Nor is it clear how the lichens propagate from one rock to another. It is indisputable, however, that these are about the hardiest eukaryotes on the planet.

Friedmann has concluded that a square meter of rock surface in the Antarctic dry valleys may harbor on average nearly a third of a kilogram of lichen. Using this value we can make a ballpark estimate of the amount of cryptoendolithic lichen biomass possible in a world with no vascular plants, assuming that cryptoendoliths could survive in a variety of climates. This is a reasonable assumption because cryptoendolithic green algae have also been described from semiarid lands and deserts in the southwestern United States.[6]

Our ballpark calculation of the amount of cryptoendolithic biomass that could be supported worldwide, given the present rock surfaces but assuming no soils or competition with plants, is 130 gigatons. This total biomass would contain about 29 gigatons of carbon (compared with between 420 and 830 gigatons of carbon sequestered in vascular plants today).[7] "Lichenland," therefore, might potentially have generated only about 5 percent of the biomass that the modern terrestrial biota creates. Lest this seem like a paltry amount, however, note that this 5 percent is greater than all the living biomass carbon in the oceans.

When did the first lichens evolve? It is not yet possible to answer this question, for the known fossil record is mute on this point. It seems, however, that the timing of lichen origination was controlled by the first appearance of the eukaryotic part of the symbiosis, the mycobiont. Phycobiont ancestors, at least of the prokaryotic variety, were probably living in scattered land habitats long before the first lichen.

Bacteria on Land

Judging from the first appearance of marine bacteria in the fossil record, which apparently formed conspicuous scums or mats on the substrate, the earliest terrestrial communities probably also formed microbial mats and crusts on moist surfaces. These mats may have included both thermophilic bacteria and cyanobacteria. The rigors of the intertidal zone, with its wildly fluctuating temperatures and salinities and periodic desiccation would have served as a stepping stone for the evolution of fully terrestrial modes of life.

As described in chapter 2, geochemical models of the carbon cycle and climate through time suggest that there may have been a strong degree of biotic enhancement of weathering even several billion years ago. Is there any direct or indirect laboratory evidence for land life several billion years old? The answer may involve studies involving the mining of precious metals. It turns out that bacteria are involved in the biotransfer and accumulation of forty or more mineral species, including gold and uranium ores. In a 1974 paper that has been described as having a "definite landmark quality," D. Hallbauer and K. Warmelo presented evidence they said confirms the presence of land eukaryotes three billion years ago. The putative fossils are directly associated with gold from the famous gold deposits of the Witwatersrand, South Africa, which are 2.5 billion years old. The fossils were originally interpreted by the authors as terrestrial fungi or lichens capable of concentrating gold and

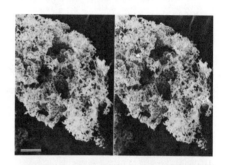

FIGURE 4.3

Bacteria replaced by gold from the Livengood District, Alaska. Mineral replacement is not an uncommon method of fossilization. This placer particle represents a fossil colony of budding bacteria. Stereoscopic pair. Scale bar = 18 microns.

FIGURE 4.4

Enlargement of the right edge of figure 4.3. Budding bacteria replaced by gold are clearly visible. Stereoscopic pair. Scale bar = 4 microns.

uranium. Along with the primary fossil material can be discerned columnar carbonaceous particles, also assumed to be fossils, that look like tiny gymnosperm cones.

Doubt has been cast, however, on the biogenic nature of the Witwatersrand structures. An experiment conducted by Preston E. Cloud and Dave Pierce, using particulate carbon and a laboratory oven, abiogenically produced "pseudomicrofossils" very similar to the putative fossils described from the Witwatersrand. On the other hand, the case for Witwatersrand fossils has been bolstered by more recent research conducted by John R. Watterson, who has found bona fide solid-gold fossils of a similar form and pattern. Placer gold particles from the Lillian Creek, Livengood District of Alaska appear to be colonies of *Pedomicrobium*-like bacteria in a stage of hyphal budding (figures 4.3–4.5); they were fossilized by replacement with gold, and the rock in which they are found is indisputably terrestrial. According to Watterson, the Livengood specimens closely resemble the stringy gold extracted from carbonaceous matter from the Witwatersrand. Because Watterson has demonstrated the widespread nature of bacterioform gold, the Witwatersrand findings demand renewed attention as possible direct evidence of terrestrial life several billion years ago. That life may not have inhabited a bare surface or soil—it may have lived in ground water or in a shallow stream.[8]

If Watterson's work does confirm the biogenic, or even eukaryotic, nature of the Witwatersrand specimens, it would fit

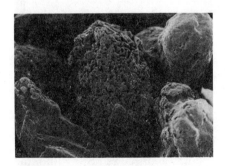

FIGURE 4.5
High-purity bacterioform placer gold particles from the south coast of the Nome area of the Seward Peninsula. Prior to examination by scanning electron microscopy, the gold sample was boiled for a few minutes in a solution of phosphoric acid. Scale bar = 40 microns.

well with the inferences pointing to an early land microbiota drawn by David Schwartzman and Tyler Volk from their modeling of geochemical and climate history. Overall, it seems likely that a significant moneran microbial biota, and perhaps even simple eukaryotes, existed on land well before the Cambrian boundary and perhaps even for the majority of earth history. But what of the earliest forms of complex life on land?

Plants Defined

Before we discuss the specifics of complex life on land, it is essential to review the main characteristics and structural features of plants. We are taxonomically rigorous in our use of the term *plant*. Plants are all eukaryotes, members of the kingdom Plantae, and they all have undulipodia, mitochondria, and chloroplasts. More important for our purposes here, however, plants can be distinguished by their (1) multicellularity, and particularly, multicellular reproductive organs, (2) photosynthetic capabilities, and (3) their development from an embryo. For example, lichens are not plants because they lack an embryo. Because of this, lichens were derisively referred to earlier in this century as "evolutionary dead ends." All plants share a common ancestor that must have been some sort of aquatic green alga (a chlorophyte).

Plants are divided into two groups, the vascular plants and the nonvascular plants, distinguished by the presence or absence of a system of vessels for transporting fluids. Nonvascular plants of the living flora include mosses and liverworts; they are members of a group known as the bryophytes. Vascular plants include the conspicuous gymnosperms (conifers) and angiosperms (flowering plants), as well as the more ancient sphenophytes (horsetails) and lycophytes (clubmosses and related forms). They possess important physiological structures adapted for the transport of fluids and solutes (sap) within the plants' bodies. (Nonvascular plants largely lack these special plumbing abilities.) With the help of tubular cells called tracheids, and other transport cells, vascular plants are able to carry mineral nutrients and water upward (through tracheids in the xylem) and to carry solute nutrients formed by photosynthesis downward (via sieve tube cells in the phloem). The xylem and phloem systems are not directly linked, but a fair amount of leakage does occur between them.

Plants developed a lifestyle called alternation of generations. The lifestyle alternates between a diploid stage and a haploid stage. The diploid stage (also referred to as the sporophyte generation because this stage produces spores), has twice as many chromosomes as the haploid stage. Through the process of meiosis, or chromosome reduction (which halves the number of chromosomes), the diploid generation produces a special set of cells dedicated to propagation. These haploid cells become spores, which grow and develop into haploid plants. The haploid plants represent the gametophyte generation because, in completing the circle of the plant life cycle, they will eventually produce male and female gametes, which fuse their chromosomes via plant sex to begin the development of a fully diploid sporophyte plant.

The sporophyte and gametophyte generations belonging to a single species often appear quite dissimilar—so different, in fact, that sporophytes and gametophytes belonging to a single species can be mistakenly identified as separate species. The sporophyte generation in vascular plants is generally the more vigorous and longer-lived of the two. The bracken and other ferns familiar to all are just the sporophyte episodes; the fully grown gametophytes may be no bigger than a fingernail. But the preeminence of the sporophyte is not always the case. The gametophytes of the modern plants *Psilotum* (the whisk fern) and *Ophioglossum* (the adder's-tongue fern) form branching, cylindrical, subterranean plants that are quite long-lived.

The Earliest Land Plants

William G. Chaloner has suggested how to search the fossil record for the oldest vascular plants. To qualify as a candidate for this distinction, a fossil plant must exhibit: (1) spores with a resistant organic wall and a distinct bursting slit; (2) tracheids (conductive tubular cells with annular thickenings; see figure 4.6); and (3) a cuticle perforated by stomates, which are pores in the plant cuticle opened and closed by guard cells. Figure 4.7 depicts the turgid, pores-open state of the guard cells in contrast with the flaccid, pores-closed state. Chaloner's criteria are relatively straightforward and practical, since they rely on recognition of well-known features of vascular plants.[9]

Recognition of early fossil nonvascular plants, such as bryophytes, is less straightforward. An approximately one-billion-year-old fossil called *Longfengshania* (figure 4.8) has been re-

FIGURE 4.6

Cross section through geranium leaf, showing transverse sections of tracheids (conductive cells) and hairs on leaf surface. Width of photomicrograph is 778 microns.

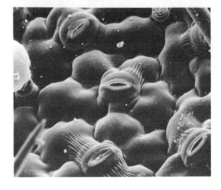

FIGURE 4.7

Stomates formed by paired guard cells on the surface of a geranium leaf. These structures admit air into the inner part of the leaf when open, and close when the plant needs to conserve moisture. Width of photomicrograph is 110 microns.

cently interpreted as a land plant with possible bryophyte affinities. First described as a marine carbonaceous megafossil of an alga, *Longfengshania* occurs in the Qingbaikou System in northwestern Hebei, China. It is a round-to-oblong structure with a short, pointed appendage called a stipe.[10]

It has been argued that this fossil has many features in common with living liverworts and mosses, as well as with the middle Paleozoic fossil bryophytes *Torticaulis* and *Sporogonites*. These arguments are based on the overall morphological similarity of *Longfengshania* to the Devonian sporophytes, and on the interpretation of a common mass at the center of a bundle of *Longfengshania* specimens (in the Little Dal Group of Canada) as an attached gametophyte. But considering the aquatic habitat indicated by the geology of the surrounding matrix, it seems unlikely that this Proterozoic fossil was indeed a bryophyte.

Tantalizing hints of Proterozoic land life occur after about 590 to 610 million years ago. Putative plant spore tetrads with trilete scars (figure 4.9) were extracted from borehole samples

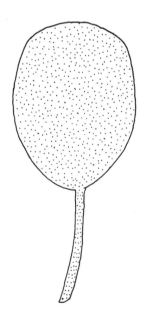

FIGURE 4.8

Longfengshania, a puzzling, paddle-shaped fossil eukaryote of the Proterozoic. The elongate "tail" is called a stipe. Length of specimen is 1.4 cm.

drilled in apparently marine sedimentary rocks in the Kotlin-skaya Suite in Latvia.[11] The spores are late Vendian in age, which puts them just before the Cambrian burst of skeleton-forming marine animals, but there is controversy as to whether or not they represent the spores of land organisms, blown out to sea.[12] Like the discussion surrounding *Longfengshania*, these spores and other suggestions of plant life on land before the Cambrian remain inconclusive.

The marine and terrestrial prelude to our main topic now set forth (with all of its frustrating and tantalizing uncertain-ties), we will now to turn to the evolutionary challenges that brought about Hypersea.

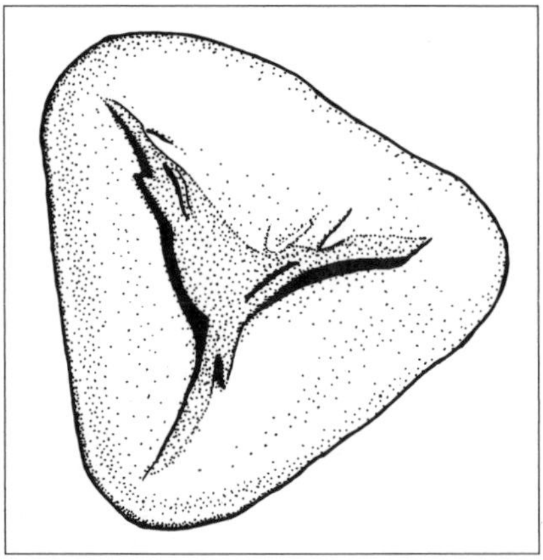

FIGURE 4.9

Putative Proterozoic spore from Latvia. Note presumed trilete marking. Greatest dimension of spore is approximately 7 microns.

FIVE

Exploiting Osmosis for Hypersea Physiology

Study of the adaptations of plants and animals is called ecophysiology. Ecophysiologists try to understand how various parts of a creature's body help it survive in its particular environment. There is even a field of study that might be called anti-ecophysiology, in which scientists argue that particular traits of organisms (such as the contours of the internal parts of plant stems) are actually not adaptations directly related to the survival of the organism, but rather are constructional artifacts, byproducts of growth patterns.[1]

It is certainly true that not all recognizable features of organisms help them survive. For instance, the impression made by a leaf on the back of the next youngest leaf while the two were pressed together in the bud (an effect easily recognized in the century plant *Agave*) is not a pattern that, as far as we can tell, helps perpetuate the plant species. The leaf-back pattern is thus a fabricational feature rather than an adaptive trait. Certainly many features of plants can be correctly interpreted as adaptations for survival, but one must exercise care lest adaptationist interpretations be overextended.

Caveats about anti-ecophysiology aside for the moment, we will now analyze some of the features of early colonists of terrestrial habitats that seem to have helped them survive. With our Hypersea concept, the conventional notion of ecophysiology is given a twist because insights can be gained from looking at not just the physiologies of particular organisms on land, but at an extended physiology that the terrestrial ecosystem as a whole makes for itself.

Stepwise Evolution of Ecophysiology in Plants

Viewed from the perspective of a marine animal or green alga, plantless terrestrial habitats would have seemed dauntingly inhospitable and alien, almost like the surface of another planet. Life on land for plants requires novel adaptations to deal with four main challenges. The first is facilitation of spore dispersal and cross-fertilization. The second is regulation of air infiltration so as to enable the intake of sufficient carbon dioxide for photosynthesis while resisting excessive water loss to the atmosphere. Third is to obtain nutrients and excrete wastes. Fourth is the transport of water and dissolved substances within the plant.[2]

These challenges were experienced early by multicellular green algae. Some forms lived in transient pools of fresh water or on moist soil surfaces, and so must have already been adapted to desiccating conditions. Early land plants probably started out with a passive water economy, in other words, an economy dependent on rain and simple capillary action to supply needed water. Bryophytes retain this reliance to this day.

Some lineages of plants evolved further, moving toward an active water economy by developing roots or rootlike structures (rhizoids) to absorb water, tubes (tracheids) for internal conduction, and mechanical tricks to save water (such as stomatal regulation). Rootlet absorption of water and tubular transport of internal water may have been the first steps taken toward active water regulation in early plants.

Ecophysiological considerations not only help us identify the structural and physiological advances that would have been necessary for fuller use of the terrestrial environment; they may even suggest a particular chronological order for the invention of those adaptations. Plausibly the first significant new invention for terrestrial life was alternation of generations, in which simple fusion of gametes (as in animals and many unicellular photosynthetic eukaryotes) was superseded by alternation of a haploid gametophyte and a diploid sporophyte (spore-producing) generation. From the perspective of a land organism, spores are a big improvement over free-swimming gametes for the simple reason that spores can be dispersed by the wind, and hence propagate the species without having to be constantly immersed in water.

The newly evolved sporophyte stage presented, in turn, a new problem: how, specifically, to disperse the spores. So the second invention for terrestrial existence was probably an effective means for dispersing the spores—notably a sporophyte body that could raise itself well above the ground in order to take advantage of winds. But exposure to the breeze exacerbates the problems of desiccation. The third evolutionary invention of land plants was probably therefore development of a waxy cuticle to prevent drying.

If it was advantageous to get above the ground, it was probably advantageous to get even higher, but ceilings on water-transport capabilities would have to be surmounted. And so the fourth invention was likely the tracheid-rich conductive tissues called xylem. Tracheids are long cells that sacrifice themselves for the life of the plant as a whole; after a tracheid cell dies, its cytoplasm is replaced by a continuous stream of fluid. Evolution of tracheids was accompanied by the evolution of strengthening materials, like lignin. Lignin is an extremely complex hydrocarbon that comes in many varieties; it is a core constituent of wood. Together, the invention of tracheids and lignin made it possible for vascular plants to grow considerably taller than their nonvascular counterparts.

The fifth essential novelty was the invention of intercellular gas spaces to enhance the exchange of carbon dioxide and oxygen. An active method of regulating gas exchange (O_2 and CO_2 in; O_2 out) would probably have been the sixth adaptation. This was accomplished by the invention of guard cells (refer back to figure 4.7) at the edges of the pores (stomates). A marvel of biotic engineering, the guard cells come in pairs on each stomate. When the plant is flush with water, these guard cells automatically become turgid like the rest of the plant, and their fullness pulls the stomates open. But during times of water stress their flaccidness closes the stomates, thus preventing further water loss from the intercellular gas spaces.

The stratigraphic first occurrences of these six adaptational novelties are consistent with this particular stepwise perspective on ecophysiological evolution. (They do, however, appear in a fairly rapid series, so one cannot be absolutely certain of the order.) Several of the steps appear to have been taken independently by different groups of plants during the Devonian, and many of these plants appear to have reached the same evolutionary grade at about the same time. Later, a seventh step is taken: great increase in the maximum size of the sporophyte. This, too, occurs independently in separate lineages of Devonian plants.

Osmotic Challenges For Animals

Although all life on land must find some adequate solution to the problem of support in air, to the challenges of dispersing propagules, and to the rigors of pronounced temperature fluctuations, far and away the most important problem for all organisms on land involves retention of water. For animals, the amount of water vapor in air is usually more critical than any temperature extremes. Air is normally below its saturation vapor pressure, which means that water will be lost through unprotected membranes. If water-bearing air is cooled enough during the diurnal cycle, the saturation vapor pressure will drop sufficiently for dew to precipitate out of the cooled air. For some desert animals, dew is virtually the only water source.

There are two routes onto land for animals. Land animals

can evolve either directly from marine ancestors or from fresh-water intermediates. The blood chemistries of land animals can indicate which route was taken.

A solute is a dissolved substance like a salt. Osmotic pressure is a measure of the amount of solutes in a fluid; high osmotic pressures denote a salty fluid, and low osmotic pressures indicate a more dilute fluid. Osmosis is the diffusion of a solvent such as water through a membrane impermeable to the solute; higher concentrations of solute on one side of the membrane create a physical pull on the solvent, causing it to move to the saltier side of the membrane. This is why fingers and toes wrinkle in the bathtub—fresh water from the bath moves into the skin (which acts here as a differentially permeable membrane) and expands the skin, causing it to temporarily swell.

Land animals with the highest blood osmotic pressure (salti-est blood) are those that have evolved directly from marine ancestors. Those with lower osmotic pressure are in general those that evolved from freshwater ancestors. Of course, all land animals ultimately had marine ancestors; but some lineages passed through freshwater environments en route to land. These particular lineages have more dilute blood as a result. For example, the snail *Pomatias elegans* has very high osmotic pressure for a land snail, about 25 percent of the osmotic pressure of normal seawater. This high internal osmotic pressure indicates that *Pomatias* is descended directly from a marine ancestor. In contrast, land snails belonging to the family Cyclo-phoridae have blood concentrations that are less than 10 percent that of seawater, indicating that their immediate ancestors were freshwater aquatic forms.

Land snails like *Pomatias* with high osmotic pressures are extremely tolerant of desiccation and the concomitant rise in the osmotic pressure of their blood. Osmoconformers such as these snails passively tolerate environmental changes, leaving each cell in the snail's body to cope with the osmotic stress.

Land snails that have low osmotic pressures, such as the

cyclophorids, are osmoregulators, which means that they use energy to maintain the osmotic pressure of their body fluid within a certain range. Their osmotic pressure is maintained by an active kidney, which produces copious amounts of dilute urine. These cyclophorid land snails are so similar to their freshwater relatives (also members of the family Cyclophoridae) that they have been referred to as aquatic snails living on land. The contrast with osmoconformers such as *Pomatias* (which has almost 400 percent higher osmotic pressure in its body fluids) is striking. Because of the need to produce dilute urine, cyclophorid land snails require secure access to fresh water. They are thus restricted to relatively humid environments. Ironically, the snails that skipped the transitional freshwater stage have an osmotic advantage in dry land environments.[3]

Freshwater protoctists and sponges get rid of excess water using contractile vacuoles, thus maintaining relatively high internal osmotic pressure. In more complex animals (metazoans), the energy required for regulating cellular ionic components is conserved by setting up the internal environment we call blood. Special organs in metazoans are dedicated to internal osmoregulation. In flatworms, for instance, this is accomplished with flame cells: ducts lined with cilia formed by infoldings of the animal's skin. The flame cell is a blind-end tube; cleansed body fluid is drawn as a filtrate from the body fluids that enter into the inner cavity, and are then swept out through the duct by ciliary beating.

Larger animals have kidneys, collections of tubules used to modify blood filtrate primarily for regulating the salt content of the animal's body fluids. On its "upstream" end, the tubule has a glomerulus, which produces a very pure filtrate of the blood—that means, only the plasma with the impurities screened out. The filtrate then advances "downstream" through the tubule, which makes a hairpin turn called the loop of Henle. The tubule straightens out again and ends in a collecting duct. The greater the amount of osmoregulation required by a given animal, the longer the loop of Henle needs to be.

Marine invertebrates and sharks have blood with a salt concentration similar to that of seawater, or somewhat over 1,000 mOsm (mOsm is an abbreviation for milliOsmole, a unit used to measure osmotic pressure). Terrestrial insects and vertebrates, as well as marine vertebrates, have a salt concentration in their body fluids a bit less than half that of seawater—between 400 and 500 mOsm. For comparison, *Pomatias* has an internal salt composition of 254 mOsm, and the osmoregulator cyclophorid land snails have a value of about 68 mOsm.

Water regulation in insects and some other small land animals is aided by the impermeability of the cuticle. Using a mechanism which is still poorly understood, these organisms are actually able to take up water from unsaturated air that is below the saturation vapor pressure.

Plant Exploitation of Osmosis

For plants, osmosis is more of an opportunity than a challenge. For this reason, plants have collectively evolved into the primary vessels of the circulatory system of Hypersea. Following is a brief discussion of how plants move fluids from the ground up into their leaves.[4]

Recall the concept of osmotic pressure, the measure of the amount of solutes in a fluid. Plant body fluid, like animal body fluid, contains solutes and thus has a higher osmotic pressure than the relatively fresh water in soil. Roots planted in soil tend to absorb this water, and they would swell up and eventually burst if not for one key fact of plant biology—plant cells have rigid cell walls.

Rigid cell walls are important because they can stop the influx of fresh water into a cell without the cell having to expend any energy to pump the water out by osmoregulation (a significant cost incurred by freshwater animals) or other means.

This is so because of a phenomenon called pressure potential. If a sack that is formed of a membrane permeable to water but not to solute, such as dissolved salt, is filled with salty water and then placed in fresh water, fresh water will move into the sack to dilute the solute within. However, if pressure is applied to the sack (say, by squeezing from the outside), fresh water will not flow in to dilute the salty water. It is the rigidity of the plant cell wall that provides the pressure to keep excess water out of the cell's cytoplasm, by counteracting the osmotic pressure tending to draw water into the cell. Water regulation in plants is thus partly accomplished by the squeeze imparted to the watery contents of cells by rigid wall pressure.

How do plants use osmotic pressure and pressure potential to move water up to the aerial parts of the plant? Since ground water has less solute than does plant fluid, the plant absorbs water. The amount absorbed is regulated in part by pressure exerted on the fluid by rigid cells walls; once again, if it were not for this rigidity the root cells would swell and burst. How do the roots maintain a higher solute content than that of ground water? Root fungi seem to help here by actively absorbing mineral nutrients (solutes) from soil minerals such as clays. More importantly, however, is that water is continually being lost from the upper tissues of the plant via transpiration of water from the leaves. Evapotranspiration is actually the main factor enabling roots to take in water. This evaporative loss leads to an overall increase in the solute content of the fluids of the plant, and this moisture loss is osmotically felt in the roots. The root cells cannot swell, thanks to their rigid cell walls; but instead of stopping water influx altogether, the rigid walls capture water by allowing the water to pass through them. Since the root cells are in osmotic communication with the vascular system of the core of the root, the incoming water is transmitted to the vascular system for a ride up the xylem to replenish the thirsty leaves.

The rate of transpiration is the main factor determining the rate at which water is drawn out of the soil. Evaporation of water from the wet cell walls into the intercellular gas spaces of transpiring tissue (such as a leaf interior) and out through the stomates increases the solute content of cells by making the remaining water more solute rich. This forces water to diffuse into the leaf spaces from the cytoplasm or vacuoles of the cells and also from the walls of adjacent cells that are not in direct contact with an air space. This water must be replenished from below. Upward flow rates in xylem tracheids as high as 4.5 meters per hour have been measured, and rates of about 1 meter per hour are typical of transpiring plants. Mineral nutrients are also carried to the leaves and other above-ground tissues by this flow.

Mention should be made of a particular strategy employed by plants to combat desiccation.[5] The presence of a film of fat or oil is a proven means to prevent desiccation. In the liverwort *Pellia*, the clubmoss *Lycopodium*, and the ferns *Ophioglossum* and *Botrychium*, desiccation resistance is accomplished by oil deposits in special zones of the plant which envelop branch-tip and reproductive tissue, effectively sealing off these tissues from the effects of drying. Many other plants have a thin film of oil over the surfaces of their leaves. The wonderful smell when you brush past a basil or a mint plant is caused by your disruption of this oil layer.

The ability of land plants to absorb water from soil is at the "heart" of Hypersea. Hypersea is a biogeophysiological entity composed of a circulatory system evolved by (and for) its biological components. The geophysiological impact of this system profoundly affects the surface of the earth. Here we are using the term *geophysiology* in perhaps an even more literal sense than intended by James Lovelock, who coined the term.[6]

Extending Vernadsky's Law

Vladimir I. Vernadsky (1863–1945; figure 5.1) is regarded in Russia as the predecessor of James Lovelock and Lynn Mar-

gulis, whose Gaia hypothesis of 1974 has gained wide scientific
and media attention.[7] Vernadsky is credited with founding the
science of biogeochemistry and with initiating a new theory of
the biosphere. He has recently begun to receive (long-overdue)
recognition from western scientists.[8]

Vernadsky viewed life from a planetary perspective. He
looked at how living processes affect the absorption of solar
energy, the movement of matter, and the shifting chemistries
of earth's surface. To Vernadsky, living processes had five
main planetary functions, and he viewed these functions as
powerful forces—indeed, as special kinds of geological forces.

FIGURE 5.1
Vladimir I. Vernadsky, the
founder of biogeochemistry.

The first function of living processes is energetic, such as in the absorption of solar energy in photosynthesis and the transfer of energy via the food chain. The second function is concentrative, as in the selective accumulation of elements such as calcium for shells and bones. The third is destructive, as in the fungal ability to enhance the chemical breakdown of inorganic matter. Fourth is medium-forming (the "cheese effect"), as in the transformation of the physical and chemical parameters of the local environment by the activity of bacteria or other microbes. Fifth is transportation or the transfer of matter without the help of gravity, especially in a horizontal direction.

Of typical meteorological phenomena on the earth's surface, only tornadoes and hurricanes are able to transport matter in a horizontal fashion unassisted by gravity in quantities on the same scale as the transfer of matter by living substance. This fifth main function of life has been rendered as "Vernadsky's Law": Migration of elements in the biosphere is either accomplished via the direct participation of living matter (biogenic migration) or it proceeds in an environment where the particular geochemical parameters (O_2, CO_2, H_2S, etc.) are conditioned by living matter.[9]

In other words, by the motions of their bodies or by the gases they release, living things move enough matter to be on a par with great meteorological and geological forces. Vernadsky's law has two aspects—the workings of creatures as they move, carrying chemicals with them, and the motions of the gases and other substances released by the organisms. The first aspect of Vernadsky's law involves actual transport by the motion of living organisms, such as the migration of phosphorus atoms from sea to land when shore birds leave their droppings inland. (To a Vernadskian, a seagull is phosphorus on the wing.) Vernadsky took the motions of living matter quite seriously and developed quantitative concepts of the "velocity of life" and the "pressure of life." Vernadsky even called these biological values geochemical constants, and the formulas look like they were inspired (at least in part) by the ideal gas laws of physics and chemistry.

Vernadsky calculated the velocity of life for bacteria as one mile in five seconds, or approximately the same speed as the propagation of sound waves in air. While such a "speed" for bacteria may seem incomprehensible, its derivation is straightforward[10] and the exercise is intriguing. For example, using Vernadsky's formulas for the determination of geochemical constants, one can calculate the theoretical maximum number of organisms of a given size that could inhabit the earth's surface.

The second aspect of Vernadsky's Law is the biological "launching" of matter as biogenic gas or other substances, such as when free oxygen is liberated into the atmosphere by the action of photosynthesis. The oxygen crisis of two billion years ago, first recognized by the late (and great) Preston Cloud, involves this second aspect. Cloud, an American geologist who died in 1991, rarely cited Vernadsky's work, but he may nevertheless have been influenced by Vernadsky; he was at Yale University while Vernadsky's son George was on the faculty there. Cloud may have heard of Vernadsky's work through George.

We consider the Hypersea concept as representative of a new, third aspect of Vernadsky's Law, by which life further enhances the migration of elements on Earth's surface. This third aspect recognizes the active modulation of solutes within the tissues of an organism (primarily vascular plants). In effect it is similar to a combination of the first and second aspects but really involves a new type of substance called Hypersea. By modulating their own solute concentrations and water potential, land organisms (and to a lesser extent land organisms that have returned to the sea) create a new type of life-mediated habitat on Earth.

Consider: the more that plant fluid becomes like seawater, the more fresh water it draws into and through itself, filling the "tributary" arteries and "distributary" veins of Hypersea, which can be described as a "terrestrial sea." A medium-sized tree in the tropics transpires a thousand liters of water a day. The terrestrial sea expands with every increase in the volume of tissues and sap and lymph of the creatures that constitute it, and the countless but interconnected conduits of Hypersea all participate in horizontal and vertical Vernadskian motions.

The Hypersea perspective, especially when placed in a Vernadskian worldview, is why we like to take the term "ecophysiology" quite literally when it is applied to land life. The geological import of this new ecophysiology is apparent in the exuberance of the land biota, and it is ultimately manifest in the sedimentary rock record as the genesis of coal, a rock type virtually unknown before the advent of abundant land life.

The most important Vernadskian aspect of Hypersea was touched on in chapter 1. This is the phenomenon of hypermarine upwelling. The ability of land plants to modulate nutrient availability through the upward direction of fluid flow prevents chronic nutrient starvation like that experienced by organisms in ocean surface waters. Through root uptake, transpiration, and well-engineered control of water potential, plants create for the land biota a zone of constant hypermarine upwelling.

Overall, we conclude that Vernadsky's emphasis on the horizontal transfer of living matter (aspects one and two of Vernadsky's Law) is appropriate and well taken but that it is only part of the story for the land biota. The biogeochemically most important transfer is manifest as a third aspect we propose for Vernadsky's Law: the vertical transfer of water and nutrients by hypermarine upwelling. The implications of hypermarine upwelling for the land biota will be examined in chapter 9. In the next two chapters we will first survey the evolutionary history of land plants and land animals.

SIX

Hypermarine Aspects of Plants and Fungi

Hydrofluoric acid is a deadly poison. A drop that finds its way onto human skin may not be noticed at first, but untreated it will work its way inward and attack nerve tissue and bone. Nevertheless, hydrofluoric acid is the paleobotanist's best friend.

In the search for the earliest land plants, it is used to dissolve the silty matrix of sedimentary rocks of early Paleozoic nonmarine deposits and to recover the fragmentary fossils of early land plants locked within those rocks. These fossil fragments are insoluble in hydrofluoric acid and are left behind when the rocks dissolve. Because of the typically fragmentary nature of the acid-freed fossils, their interpretation is far from straightforward. The ancient flotsam and jetsam that ends up in these acid-insoluble residues includes embryophyte spores (spores which, on the basis of their morphology, can be attributed to embryo-forming land plants), cryptospores (spores that do not clearly belong to embryophyte plants), sheets of plant cuticle, tubular elements called nematoclasts, and occasional fragments of arthropod cuticle.

Hydrofluoric acid is thus the primary tool of what is called palynology, the study of fossil pollen, spores, and other tiny bits of plants. Pollen and spores and certain plant fragments can withstand physical and chemical decomposition that other plant organs cannot; hence their importance in piecing together the history of Hypersea. But how can one be sure that the fragments came from the particular kind of plants important to the geophysiology of Hypersea, that is, vascular plants?

Clues from Scraps and Spores

Most spores of the early Paleozoic are thought to have been formed in organs called sporangia. Sporangia occur today in nonvascular plants, the mosses and liverworts. Figure 6.1 shows spores in a sporangium from the Rhynie Chert plant *Aglaophyton*.

Recognizable land plant spores include forms such as *Ambitisporites* (figure 6.2), which has a distinctive trilete scar. Trilete scars such as this, with the three linear scars oriented at 120° to one another, indicate that the spore was once part of a symmetrical cluster of four spores called a tetrad. The trilete scar shows where the four spores were once attached to one another, each spore bearing one linear scar for each one of its three sibling spores. Today, such spores are distinctive of plants. In contrast, spores with a single linear opening slit or scar are referred to as monolete spores. Monolete spores can be formed by aquatic algae as well as by plants.

Spores with trilete markings resembling trilete scars have been reported from Vendian (Proterozoic) sediments, but it is far from certain that these are indeed plant (as opposed to algal) spores. The question is, Are these marks trilete scars, or do they just resemble trilete scars? Jane Gray has made careful

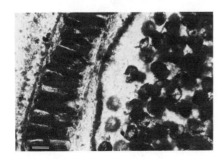

FIGURE 6.1
Sporangium of the Rhynie Chert plant *Aglaophyton*. Spores are seen on the right. Scale bar = 100 microns.

studies of early tetrads, and has convinced her skeptical paleobotanical colleagues of the existence of tetrads at least by the
Ordovician. The oldest well-dated tetrads of reasonable abundance come from middle Ordovician (Llandeilo/Llandvirn
stage) rock in Arabia. Potentially older (but still probably
Ordovician) tetrads occur beneath the Armorican Quartzite in
Spain. It is not yet known whether the producers of these spore
tetrads were semiaquatic or fully terrestrial.[1]

Cryptospores, or spores of unknown origin, come in a number of interesting varieties. A cryptospore such as *Tetrahedraletes grayii* is a sort of "permanent tetrad" formed by four
tightly fused subspores which are partially collapsed inward
toward the center of the tetrad (figure 6.3). *Nodospora retimembrana* (figure 6.4) is another cryptospore in which the tetrad
complex seems to be approaching more of a cross (or cruciate)
arrangement rather than the tetrahedral arrangement diagnostic
of plants. Other species of *Nodospora* often produce a spore-
surrounding membrane.

FIGURE 6.2

Ambitisporites sp. The prominent trilete scar indicates that this is a plant
spore. A. Width of spore is 37 microns. B. Width of spore is 27 microns.

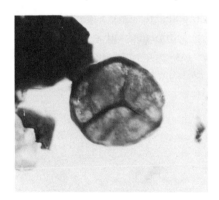
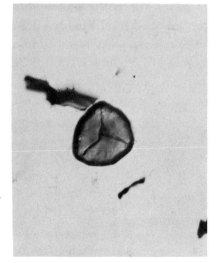

Some cryptospores are dyads, or spore pairs, such as *Dyadospora murusattenuata* (figure 6.5) and *Laevolancis divellomedium* (figure 6.6). The dyads are often recovered split apart. *Confossuspora reniforma* (figure 6.7) is a dyad with numerous circular to oblong holes or pores distributed over the entire cryptospore. *Dicryptosporites radiatus* (figure 6.8) is a distinctive dyad cryptospore with radial thickenings on proximal surfaces of individual spores.[2]

Eni,...atic sheets of cuticle, some of which are derived from land plants, are known from Ordovician to Devonian sediments. These sheets have cell-like markings on their surfaces and may have knobs or other protuberances called tubercles. These cuticular sheets are difficult to interpret, for it is not always possible to distinguish between plant cuticle and arthropod cuticle. Both types frequently occur together in acid residues drawn out of early Paleozoic rocks. Cuticular fragments can be unequivocally attributed to vascular plants only if they display preserved stomates, as is the case with cuticle from the early plant *Drepanophycus spinaeformis* which shows well-preserved stomates.

FIGURE 6.3 (Left)

Tetrahedraletes grayii. This is a spore tetrad of unknown origin, composed of four tightly fused spores. Width of spore is 29 microns.

FIGURE 6.4 (Right)

Nodospora retimembrana. This is a tetrad of spores with a cruciate arrangement. It is not certain whether *Nodospora* is from a plant. Tuscarora Formation, central Pennsylvania. Early Silurian (Llandoverian) in age. Width of spore is 46 microns.

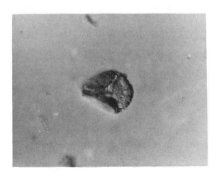

Tubular objects from acid residues, called nematoclasts, are among the most interesting of palynofloral debris. Nematoclasts can be reasonably interpreted as tracheids from the xylem of vascular plants if spiral or some other type of differential thickening or lignified banding is present. If the nematoclasts are smooth, they could belong to nonvascular organisms, such as filamentous algae or cyanobacteria. Nematoclasts from the Silurian of Pennsylvania can have spiraling, anastomosing annular, or internal fibrillar wall thickenings. When these tubes are found woven together in a weft or mat (figure 6.9), they are often assigned to the (extinct) genus *Nematothallus*. *Nematothallus* is thought to belong to an extinct group of plantlike organisms called nematophytes, as will be discussed toward the end of this chapter.

Some nematoclasts are definitely the remains of vascular plant cells. Undoubted tracheids with distinctive bordered pits (for passage of fluid between adjacent tracheid cells) are known

FIGURE 6.5 (Left)
Dyadospora murusattenuata. This is a palynomorph consisting of two spores without apertures occurring in a dyad configuration. Bloomsburg Formation, Port Clinton, Pennsylvania. Early to mid-Silurian age (possibly Wenlock). Width of spore is 28 microns.

FIGURE 6.6 (Right)
Laevolancis divellomedium. This is another dyad palynomorph of uncertain origin. Bloomsburg Formation, Snyders, Pennsylvania. Width of spore is 27 microns.

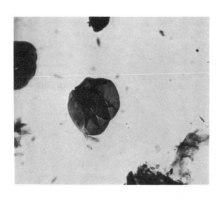

from acid residues of the fossil plant *Psilophyton charientos*. The early Devonian plant *Sennicaulis hippocrepiformis* has water-conducting cells preserved by permineralization with the minerals pyrite and limonite. These cells show a combination of the features seen in vascular plant tracheids (helical thickenings) and in nonvascular bryophyte plants (thin microporous walls of the conducting cells).[3]

According to Paul Strother, a general trend can be seen in the development of the middle Silurian palynoflora of the northern hemisphere. First came a decrease in the percentage of tetrads in the spore populations. Next was an increase in the

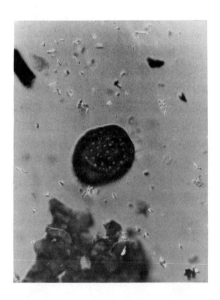

FIGURE 6.7

Confossuspora reniforma. A cryptospore dyad with numerous holes or pores distributed over the entire spore surface. Spore is approximately 27 microns in width.

FIGURE 6.8

Dicryptosporites radiatus. A spore dyad with radial thickenings on the surface of the individual spore. Spore width is 32 microns.

number and diversity of nematoclasts, followed by a slight increase in the numbers and variety of cuticlelike structures. Finally, *Ambitisporites* (figure 6.2) and similar spores became established as the dominant trilete spore type as seen in the youngest rocks.

Tracing the Evolution of Vascular Plants

Because vascular plants are the *sine qua non* of Hypersea, we now turn directly to the evolutionary history and paleobotany of these plants.

The study of Paleozoic plants has a pre-Darwinian history, dating back to the 1820s when early plants were described from the Carboniferous Coal Measures of Great Britain. In 1859 (the same year that Darwin published his *Origin of Species*) a Canadian paleontologist, J. W. Dawson, described *Psilophyton princeps* as an even earlier vascular plant of the Paleozoic (figure 6.10). His discovery was greeted harshly by critics, who accused Dawson of mistaking a Devonian aquatic alga for a vascular plant. Dawson was not vindicated until 1916, when it became apparent to paleobotanists that vascular plants existed significantly before the Pennsylvanian. With the discovery and description of the Rhynie Chert, from a site near Aberdeen,

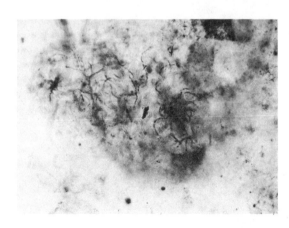

FIGURE 6.9

Nematothallus tubules. A tangled mass of tubules from the Rhynie Chert, interpreted here as belonging to *Nematothallus*.

Scotland, it became clear that most of the characteristics scientists associate with vascular land plants (upright stems, stomates, xylem and phloem, tracheids) had appeared by the early Devonian.

In 1968 Harlan Banks published a comprehensive, now-classic summary of early land plant evolution. His basic scheme, still supported by contemporary data, centered on an early dichotomy in the "trunk" of the plant family tree[4] (figure 6.11). On one side of the split are the long-extinct zosterophylls, which are usually considered to have given rise to the modern lycophytes or clubmosses. On the other side of the split are the even more ancient rhyniophytes, which are thought by some to have given rise to the trimerophytes, such as the *Psilophyton* that Dawson described (figure 6.10), and subsequently (through the trimerophytes) to all other vascular plants. Rhyniophytes are among the simplest vascular plants known from the fossil record. They consist of slender leafless stems with terminal (stem-tip) sporangia. The best known rhyniophyte genus is *Rhynia*. Rhyniophytes are characterized by having a great diversity of spore shapes.

The *Baragwanathia* Anomaly   An abiding mystery of early Paleozoic land plants involves the genus *Baragwanathia*. Known since 1935, *Baragwanathia* (figure 6.12) is found in strata of the middle part of the late Silurian (Ludlovian) of Victoria, Australia.[5] The plant was quite large for its time.

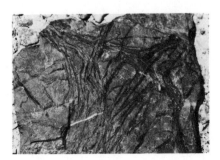

FIGURE 6.10

Psilophyton from the Devonian of New York State. Maximum width of rock in view is 80 mm. When this type of plant was first discovered, most paleobotanists refused to believe that vascular plants could have existed that long ago.

Fragments of dichotomously branching *Baragwanathia* stems are known up to 28 cm long. The stems of most specimens are 1 to 2 cm in diameter but are known to have reached 6 cm. The stems include floppy leaves only 1 mm wide and up to 4 cm long. Kidney-shaped (reniform) sporangia are present on parts of the stem, and they contain spores. The vascular system is similar to that of the Rhynie Chert plant *Asteroxylon* (which will be discussed soon), except that *Baragwanathia*'s vascular tissue is somewhat more complex.

FIGURE 6.11

Banks's dichotomy—the great divide in plant anatomy and phylogeny. Time divisions not drafted to scale.

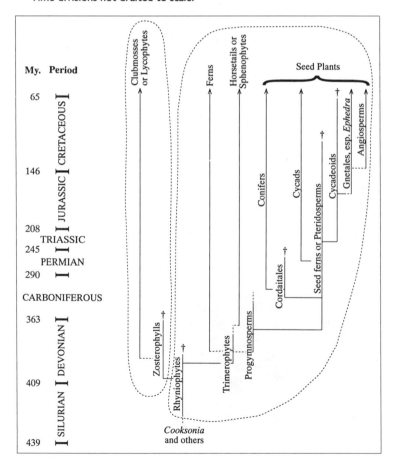

There are two unnerving aspects of *Baragwanathia*. First is that its overall shape and its stem form are strikingly similar to the living clubmoss *Lycopodium lucidulum*. Like the living lycophyte, *Baragwanathia* has true roots and leaves. Some paleobotanists are deeply disturbed to find a Silurian plant that compares so closely with a modern plant because it would be astonishing for a eukaryotic lineage to survive virtually unmodified for hundreds of millions of years. Paleobotanist Henry N. Andrews wrote in 1961 that "one may hesitate in claiming that species like *Lycopodium lucidulum* are direct descendants of this ancient Australian plant but it seems possible."

Also unnerving (to American and European paleobotanists) about the Australian *Baragwanathia* is that if it is correctly dated then it is one of the very oldest complex and relatively complete land plant fossils. This, in turn, would mean that

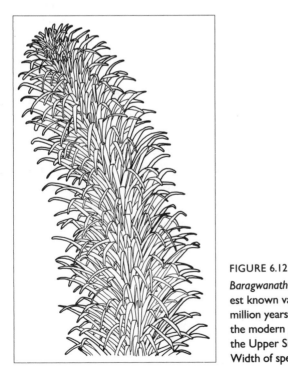

FIGURE 6.12

Baragwanathia. This fossil is one of the oldest known vascular plants. It is about 420 million years old, yet is amazingly similar to the modern genus *Lycopodium*. It is from the Upper Silurian(?) of Victoria, Australia. Width of specimen is approximately 2 cm.

plants much more complex than any Silurian forms from the northern hemisphere were present in the supercontinent Gondwana, a single giant landmass composed of today's southern continents. This *Baragwanathia* anomaly has led to the suggestion that vascular plants may have evolved first in the southern hemisphere, where most of the land was in the Silurian, and only much later found a way to spread across the sea and onto northern continents. Some paleobotanists, however, dispute the age claimed for *Baragwanathia*, feeling that the Silurian localities in which this fossil is found may be misdated because the marine fossils (interbedded with *Baragwanathia*) on which the age is based are poorly preserved.

By early Devonian times, *Baragwanathia* turns up in Canada, indicating floral interchange between Australia and the present-day northern hemisphere by this time. Trimerophytes (mostly known from the northern hemisphere) are found associated with *Baragwanathia* in the Devonian Wilson Creek Shale of Australia. *Baragwanathia*, with its sophisticated vascular system, represents a Siluro-Devonian harbinger of Hypersea that swept fairly rapidly from its southern hemispheric point of origin to the northern continents. Its early (Silurian) presence in the south leads us to speculate that Hypersea originated in or near Australia. This speculation will need to be tested by new discoveries of early plant fossils in both hemispheres.

The *Baragwanathia* mystery of complex Silurian lycophyte plants in Gondwana implies to many paleobotanists that we should expect some exciting discoveries of early land plants in lower and middle Silurian rocks—perhaps worldwide, but surely on Gondwanan continents. If *Lycopodium lucidulum* is indeed a direct descendant of *Baragwanathia*, it would be the all-time record holder for longevity; it would be the oldest "living fossil plant."

Early Vascular Plants   The oldest generally accepted fossils of vascular land plants, slightly older than the oldest known

Baragwanathia, are specimens of *Cooksonia* from the Wenlockian (Silurian) of Ireland. *Cooksonia,* best known from more
recent Silurian strata in Wales, was a tiny, delicate plant with
naked, dichotomously forked stems only a few millimeters
thick and less than ten centimeters high (figure 6.13). The
branches possessed the earliest known vascular system, preserved as a faint, thin line impression within some stems. This
impression is probably the fossil trace of conductive cells.
Reproductive structures in *Cooksonia* were mere swellings at
branch tips, containing cutenized spores,

A problem exists in confirming from the fossil itself that
Cooksonia is indeed the earliest vascular plant. Only some of
the axes (stems) belonging to *Cooksonia* show the narrow strand
of conductive cells. Based on the latest information, paleobotanist Dianne Edwards concludes that it would be improper to
assume that the cooksoniids from the Silurian Wenlockian Age
were indeed vascular plants.[6]

Dianne Edwards has succeeded in studying the spores of

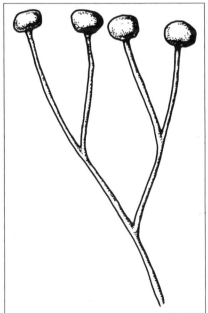

FIGURE 6.13
Cooksonia, the oldest supposed vascular
plant, may not be a vascular plant at all.
Note the knobby terminal sporangia. This
drawing reconstructs a fossil from the Upper Silurian of Wales. Distance across the
top is 1.5 cm.

later cooksoniids by breaking open fossil sporangia and examining the contents with the aid of a scanning electron microscope. Based on this research she has demonstrated that cooksoniid spores became more complex and more ornamented as time passed from the late Silurian into the Devonian.

The genus *Cooksonia* is known from Silurian to early Devonian horizons of New York State, Britain, Czechoslovakia, Russia, and North Africa. Cooksoniids may be ancestral to the group of plants called rhyniophytes (one of the main branches in Banks's scheme). *Cooksonia* is assumed to be ancestral to other plants because it is so simple and because it looks "primitive."

By late Silurian, however, the sedimentary record of the Northern Hemisphere yields undisputed fossils of early vascular plants from strata deposited in muddy shoreline environments. More complex vascular plant groups appear in the early Devonian. Many of their fossils are known from marsh sediments of ancient deltas. Forms such as *Rhynia* and the strange, bilaterally symmetric *Thrinkophyton* (figure 6.14) are thought to be descended from *Cooksonia*-like Silurian ancestors.

Rhynia is best known from the Rhynie Chert in Scotland. This genus has naked stems with ovoid sporangia at the tips, and in overall shape was somewhat similar to *Cooksonia*, except

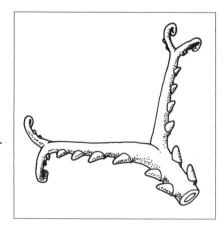

FIGURE 6.14

The Devonian zosterophyll *Thrinkophyton formosum* from Wales, Great Britain. This section of a branch has a plane of bilateral symmetry running nearly perpendicular to the page. The elliptical stele (vascular tissue) gives evidence of an internal bilateral symmetry. Such symmetry is known in no living plant. Length of branch is approximately 15 mm.

that development of the vascular system was much more elaborate. *Rhynia*'s vascular tissue was nevertheless still fairly simple, consisting of a strand of xylem surrounded by tissue that is interpreted as phloem. This vascular tissue (or stele) is also interpreted as being endarch, that is, the tracheid development progresses from the center of the first xylem to its outer edge (figure 6.15). This feature is important in establishing affinities between *Rhynia* and later plants.

The sole species of the genus is *Rhynia gwynne-vaughanii*. It seems to have had the ability to shed its terminal sporangia after spore dispersal. This may have been an adaptation to minimize water stress, and may be an evolutionary precursor to leaf dropping in later plants.

Zosterophylls Go Their Own Way Zosterophylls, exemplified by genera *Zosterophyllum* and *Thrinkophyton*, are at the base of Banks's other main branch of the geneaology of vascular plants (figure 6.11). Zosterophylls were small, branching, leafless plants. The main stem bore lateral sporangia on little stalks. *Zosterophyllum* and other zosterophylls were exarch, displaying tracheid development from the outside to the inside of the initial xylem stele.[7]

Zosterophylls are common in the Devonian. Another early Devonian zosterophyll, *Sawdonia acanthotheca*, is known for its broad, forked emergences (leaflike organs) which may be evolutionary precursors to early leaves. A recently discovered zosterophyll from the Rhynie Chert (*Trichopherophyton*) is characterized by spinose hairs.

Early Devonian floras typically contain cosmopolitan genera such as *Zosterophyllum* and *Baragwanathia*. Some localities, such as the Posongchong Formation of the Yunnan Province, China, exhibit a high proportion of endemic fossil plants having new combinations of characters not seen in previously studied finds.[8] Recent studies of this endemism of plant taxa with "hybrid" traits have suggested that the endemism is the

result of provincialism (that is, geographic isolation from genetically related organisms) in early Devonian floras.

The mixture of morphologies in single specimens, however,
could equally be due to the tendency of organisms participating
in the early stages of an evolutionary radiation to show combi-

FIGURE 6.15

Endarch versus exarch. Simplified sketch showing endarch (A) and exarch
(B) steles. The exarch lineage is circled on the left of figure 6.11 (zosterophylls and lycophytes); all other plants are of the endarch lineage and appear on the right in figure 6.11. The endarch pattern may have prevailed
because it allowed the evolution of thick wood.

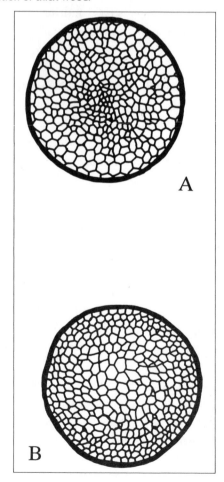

nations of morphological characters not seen in subsequent, descendant biotas. This generalization is equally (if not more) true for the animals of the Cambrian evolutionary explosion. No one is sure why this generalization holds, but it may have something to do with weeding out of less successful character combinations by natural selection or simply bad luck during times of extinction. In any case, by the end of the Devonian, the "bushy" plant family tree had been heavily pruned back to a small but vigorous sapling with a split trunk based on the exarch versus endarch pattern of xylem growth.

As the flow of geologic time passes from the early to the middle Devonian, the provinciality and limited floral exchange of the early Devonian gives way to a dramatic floral interchange. This interchange is typified by the cosmopolitan extent of *Retispora lepidophyta,* a distinctive trilete-scarred spore. This late Devonian spore occurs in North America, Europe, Asia, Africa, South America, and Australia. The appearance and spread of such cosmopolitan forms has been attributed to the shrinking of oceanic barriers as a result of subduction of oceanic crust and increasing proximity of continental blocks. By the late Paleozoic, there would be no remaining oceanic barriers to floral interchange because the converging continents would come to form a single continent—Pangea.

The exarch lineage has been heavily pruned back with regard to diversity; today it is represented only by the clubmosses. Contrast this with the endarchs—which include ferns, horsetails, conifers, and the vast empire of flowering plants. Perhaps the reason for this imbalance in the modern diversities has to do with limitations inherent in exarch forms of vascularity.

For example, consider the late Paleozoic arborescent lycophytes. These plants reached heights exceeding fifty meters and proved to be important coal formers. Despite the large size, the stems of arborescent lycophytes seem "cheaply constructed" compared to those of modern woody trees. The main support tissue in arborescent lycophytes was a thick bark, and the actual amount of wood formed by these plants was small. The wood (xylem) that was present seems to have been efficient at conducting water upward, but unlike modern woody plants there appears to have been no return flow. No phloem is seen in the stems or leaves of the treelike lycophytes, which leaves paleobotanists wondering how foodstuffs derived from photosynthesis were able to get to the roots. One suggestion is that since the roots were very shallow, they may have been able to photosynthesize independently by means of leaflike rootlets that were exposed to air. If such was the case, the sugars produced by photosynthesis would have been used very locally rather than being transported long distances along a stem.

Large exarch plants later proved to be an improvement over the arborescent lycophyte design, principally because all woody exarchs have well-developed phloem, plus the ability to make thick wood within the stem interior. Most modern woody plants are endarchs because only endarchs are able to build thick, lignin-bound palisades of conducting cells, forming a structural stem support of tremendous strength. This endarch wood, which contributes so much biomass to Hypersea, has apparently proved to be the superior mode of stem stiffening over the course of geologic time.

Harlan Banks's distinction (figure 6.11) between endarch *Rhynia*-type plants and exarch *Zosterophyllum*-types thus represents a basic division for subsequent plant evolution. The exarch zosterophylls reputedly gave rise to the lycophytes, while the endarch rhyniophytes ultimately gave rise to all the other vascular plants. The zosterophyll-lycophyte lineage has remained a separate lineage to the present, completely apart from the endarch lineage which evolved forms such as the trimerophytes which, in turn, gave rise to everything from horsetails to oak trees. There is some question as to whether or not lycophytes were descended directly from zosterophylls, or from another early exarch plant group. Reports of zosterophylls found with *Baragwanathia* indicate that zosterophylls were a very ancient group, and are consistent with the hypothesis that zosterophylls were indeed ancestral to lycophytes.

Zosterophylls were very peculiar plants in a number of ways. First, they are the only major group of plants known to have gone extinct during a mass extinction event. Zosterophylls suddenly vanished in the Frasnian Age of the late Devonian, which coincides with major extinctions in the marine biota. Zosterophylls are also peculiar in that some species had both an internal and an external bilateral symmetry expressed in the branches.[9] This type of symmetry is very unusual for a plant. The odd bilateral symmetry holds for both the external (branches, sporangia) and internal anatomy (figure 6.14). An early form, *Zosterophyllum myretonianum*, had normal, radial symmetry with sporangia spiraling up the stem. But later forms such as *Thrinkophyton formosum* had clear bilateral symmetry, seen externally in the geometry of the branching and the placement of sporangia and internally in the elliptical shape of its vascular strand. An elliptical stele is seen in no other type of plant, living or extinct. The bilaterally symmetric forms represented an evolutionary dead end—perhaps because there was something intrinsically maladaptive about a symmetrical plant or perhaps just because of bad luck. Because they vanished during a planetary mass extinction, we will probably never be able to judge.

Some zosterophylls have terminate axes (buds or sporangia at the growing tip of the plant); others have nonterminate axes (no bud or sporangia at the tip of the plant). Terminate axes seem to be associated with radial symmetry of the stem itself, as in *Zosterophyllum myretonianum*, whereas nonterminate axes are associated with elliptical steles as in *Thrinkophyton formosum*. It is the latter, bilaterally symmetric forms that failed to leave descendants. The spiral zosterophylls, as discussed next, may have left a living legacy.

Knowledge of Paleozoic terrestrial ecosystems is hampered by the lack of living zosterophylls; they were major components

of mid-Paleozoic ecosystems. *Trichopherophyton*, the newly discovered zosterophyll from the Rhynie Chert, is characterized
by its unicellular spinose hairs, exarch xylem, and laterally
attached, marginally splitting sporangia (found also in many
other zosterophylls). Zosterophylls are abundant as well in
other Devonian strata.

Although they went extinct in the Devonian, some zosterophylls may have left descendants. The clubmosses or lycophytes
may be descended from the spiral zosterophylls. The main
alternative to this hypothesis is the suggestion that lycophytes
evolved directly from *Baragwanathia* or a similar ancestral
plant. If the latter hypothesis is true, then we are presented
with an intriguing situation, since *Baragwanathia* itself is a
lycophyte. One can imagine a scenario in which an ancient
zosterophyll was ancestral to *Baragwanathia*, making zosterophylls the very earliest vascular plants—an interesting possibility, since it might then be possible to explain zosterophyll
bilateral symmetry as an early biotic experiment in the exploration of the vascular plant life style. Future fossil discoveries
may resolve the question.

Modern, creeping species of *Lycopodium* have a growth
habit that closely resembles that of the extinct, ground-creeping
zosterophyll *Kaulangiophyton akantha* from early Devonian
rocks in northern Maine. Figure 6.16 shows the Nodding
Clubmoss or Wawaeiole (*Lycopodium cernuum*), a creeping lycophyte that we observed growing across a road at Kilauea
volcano in Hawaii. (That particular road had been abandoned
and turned into a foot path after being destroyed by a volcano-
related earthquake in 1983.) *L. cernuum* is a cosmopolitan
species, occurring in locales as diverse as Malaysia, Japan, and
the American south. In Hawaii, *L. cernuum* creeps along the
ground for great distances, and at intervals sends off upright
stems up to a meter high. These stems, like those on the
Devonian zosterophylls, bear short, forking branches. The
entire plant, except the rhizomes, is covered with abundant
awl-shaped leaves.

One major difference between zosterophylls and *L. cernuum* is that the former had sporangia in separate bulges on the sides or tip of the stem, whereas the sporangia of the latter are enclosed by leaves (figure 6.17). In spite of minor morphological differences, it appears that the ecological niche occupied by these related plants has been held by zosterophylls or zosterophyll relatives continuously since the early Devonian, and that even today, *L. cernuum* can hold its own and even expand its geographic range in a world of aggressive, modern grasses. *L. cernuum* is found at Kilauea in great numbers, successfully coexisting with exotic weedy grasses.

Speaking of successful coexistence, recent research has revealed infestations of fungi within living, healthy specimens of

FIGURE 6.16
Lycopodium cernuum, the Nodding Clubmoss or Wawaeiole. Wild specimen shown here was photographed on the rim of Kilauea Crater in Hawaii. Note terminal sporangia; width of photograph is about 15 cm.

FIGURE 6.17
Lateral creeping axis of *Lycopodium cernuum* striking out across the pavement. Lycopodium is similar to the early Devonian zosterophyll *Kaulangiophyton.* Dime for scale is 18 mm in diamter.

L. cernuum.[10] The fungal hyphae infect both the gametophyte (in this case, a surface-living form) and the sporophyte generations of the clubmoss. The hyphae growing inside some of the plant cells are thick walled, and form coils encapsulated by a thin layer of clubmoss cell wall material. In other cells the fungal hyphae are thinner walled and are differentiated into tree- or bush-shaped arbuscles. They form by multiple bifurcations at the tip of a hyphal branch and are thought to play a role in the hypermarine transfer of nutrients between the symbiont fungus and the host plant. Arbuscles turn out to be very important for the phenomenon we call hypermarine upwelling; as we will show, they are best developed in the mycorrhizae of flowering plants.

It is possible that the success of the modern clubmoss *L. cernuum* is related to the fungi that are so heavily involved in both its gametophyte and sporophyte life stages. The reasons for this will be clearer when we discuss, later in this chapter, both the clubmoss fungi and the fungal symbionts of vigorous tall fescue, a common pasture grass. Hypermarine interactions as seen in *L. cernuum* were quite probably present in ancient exarch plants as well. We predict that fossils of *Baragwanathia* will be shown to contain similar fungi.

A leaf is a photosynthetic organ, attached to a plant's stem, that has its own internal vascular system connected to that of the stem. The earliest leaves were awl- or strap-shaped; they are called microphylls. There are two schools of thought regarding the evolution of the earliest leaves (figure 6.18). One idea, the telomic theory, sees leaves as developing by the reduction (overtopping) and fusion of lateral branch systems or telomes. Overtopping means that the elongation of branch tips is limited. This keeps the tips close together, allowing them to be connected by webbing that forms, together with the shortened branches, leaflike organs. In these overtopped and webbed branches, vascularization is primarily channeled to a single or a few vein branches. In short, telome theory says that leaves developed as webbing between the tiny twigs.

Leaves in the endarch lineages probably developed via this path.

The second theory for the development of microphylls is called enation theory. This view sees leaves as developing from unorganized lumps of tissue called enations or emergences. These photosynthetic lumps projected from the stem like warts and are regarded by enation theorists as the precursors of leaves. Enations were not leaves to begin with, however, because they lacked vascular tissue. According to enation theory, the development of microphylls in the exarch lineage proceeded via progressive vascularization of (i.e., extension of vascular tissues into) emergences. Some emergences were forked before they were vascularized, as in the forked emergences of the zosterophyll *Sawdonia acanthotheca*.[11]

Patricia G. Gensel has shown that a telomic ancestry for forked-leaf lycophytes is unlikely because of a lack of fossils of

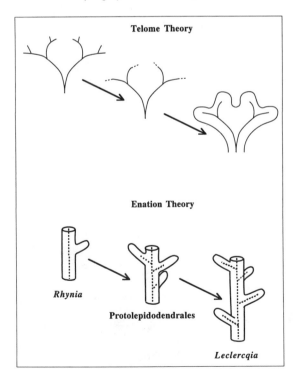

FIGURE 6.18
Telome versus enation theory. *Top:* Development of leaves by telomation and overtopping of branches. This is apparently the evolutionary progression that created foliage in endarch plants. *Bottom:* Development of leaves by progressive vascularization of enations or emergences. This seems to be the evolutionary sequence of foliage development followed by the exarch lineages. (*Rhynia,* Protolepidodendrales, and *Leclercqia* as shown represent evolutionary grades and are not necessarily representative of a direct ancestor-descendant lineage.)

possible ancestors with appropriately "overtoppable" branches. In other words, plausible intermediate forms in a lineage leading from a cooksoniid to an early Devonian lycophyte like *Leclercqia* (figure 6.19), as would be predicted by telome theory, have not been found. We agree with Gensel's assessment, and thus the vascularization of enations rather than the webbing of shortened branches is the hypothesis we favor for explaining the origin of leaves in the lycophyte lineage.

These enations or emergences evolved (either via a *Baragwanathia*-like or zosterophyll ancestor) into the spirally arranged, vascularized microphylls of the clubmosses or lycophytes. *Asteroxylon* (figure 6.20), a Rhynie chert plant associated with *Rhynia* but of significantly more complex construction, possessed microphylls of the enation type that had a definite covering or cuticle. Middle Devonian herbaceous lycophytes also sported enation leaves. This simple leaf pattern of the lycophyte lineage carries over to the giant clubmosses and the arborescent lycophytes of the Coal Age. Perhaps the best known of all fossil plants, lycophyte trees were entirely unlike modern angiosperm trees. Because of their simple leaves and lack of thick woody tissue, these lycophytes were more like gigantic versions of modern herbaceous weeds.

Endarchs Branch Out   Meanwhile, leaf development was also taking place in the endarchs—the branch of the evolutionary tree that ultimately would prove most successful. In the

FIGURE 6.19

Leclercqia from Devonian strata near Gilboa, New York. The lack of *Leclerqia* ancestors with overtoppable branches suggests that its leaves developed by the vascularization of enations. Length of fossil in view is 74 mm.

endarch or trimerophyte lineage, leaves were developed by the process of overtopping. In most of the endarchs, terminal branches or telomes were variously reduced or truncated (over-topped), webbed, planated (placed in the same plane), and fused to produce a planar photosynthetic organ. *Rhynia*'s descendants in the middle Devonian, the trimerophytes such as Dawson's *Psilophyton* (figure 6.10), still had no leaves but were more robust and had begun to show branching triple points and uneven dichotomies—precursors of telome development. In fact, the process was well underway already in the early Devonian, as indicated by Gensel's discovery in 1984 of a still unnamed trimerophyte from Quebec with a level of stem-leaf differentiation significantly more complex than known for any other taxon of this age.[12]

By the close of the Devonian, a substantial diversity of woody land plants were part of the landscape. Wood, also called secondary xylem, is formed as stacked palisades of conductive cells (tracheids) that continue to serve the plant after

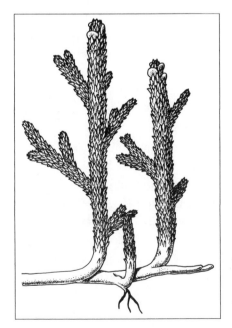

FIGURE 6.20
Asteroxylon, a spore-bearing plant from the Rhynie Chert deposit. Sporangia are the lens-shaped objects near the tops of the branches. Height of plant is approximately 15 cm.

their death. During the life of the cell, lignified bands (like the steel straps around a wooden barrel) form on the tracheid. These lignin bands keep the deceased cell from collapsing, which it might otherwise do when it loses its cell contents and begins to transmit the flow of vascular fluid. Columns of such cells give a stem great mechanical support and strength. Among the first wood-producing plants were the endarch progymnosperms, shrub- or small tree-like plants with solid, lobed steles (vascular bundles of xylem and phloem). Progymnosperm photosynthesis was accomplished in frondlike telomes.

Archaeopteris, a progymnosperm, is one of the earliest trees and is perhaps the oldest plant with well-developed, flat spreading leaves. As such, it was the first plant capable of creating deep shade. The petrified wood of *Archaeopteris* is a common fossil. *Archaeopteris* also had what is called a bifacial cambium, or active growing zone, in which new phloem vessels are formed in an outward direction (forming the vascular tissue that brings photosynthate down to the roots) and new xylem vessels on the inner face of the cambium, forming new sapwood. To envision this, imagine a cylinder like the cardboard core of a roll of paper towels. If this cylinder is a bifacial cambium, phloem cells will bud off from the outer surface (outer face), and xylem cells will bud inward from the inner surface or face of the tube. The cylinder also increases in diameter with continued growth; this is how trees increase their girth with age.

As in all trees today, the bifacial cambium causes layers of old xylem (wood) to form inside the tree trunk, while old phloem (outer bark) is spalled off as the entire trunk enlarges. In a very real sense, trees ever since *Archaeopteris* regrow themselves with each new season. Last year's vascular tissues are no longer alive this year. This is so because of the steady cell-forming activity of the bifacial cambium. Older xylem cells can in some woody plants continue to transport water for a number of years.

Unlike the stems and trunks, the roots of all vascular plants are exarch. *Archaeopteris* seems to have been the first plant with deep, penetrating roots. The development of these roots eventually led to a pronounced increase in the underground extent of most plants.

Late Devonian exarch plants are typified by the arborescent lycophytes. Quite treelike in outward appearance, these plants had spirally arranged, dense leaves in which phloem was absent. As noted earlier, trunk support was provided by a thick bark formed by stem cortex cells. The center of the stem was filled with thin-walled, living cells separated by air spaces. One can tell that the plants are exarch because the largest cells in the stem are right at its core.

Arborescent lycophyte genera such as *Lepidodendron* (figures 6.21 and 6.22), *Lepidophloios* (figure 6.23), and *Sigillaria*

FIGURE 6.21 (Left)
Lepidodendron aculaetum, an arborescent lycophyte from Locust Dale, Pennsylvania. In this, as in all the arborescent lycophytes, the repeating patterns on the surfaces of the trunks are primitive leaf scars. Specimen width is 17.2 cm.

FIGURE 6.22 (Right)
Lepidodendron sp., a Pennsylvanian arborescent lycophyte. Long axis of each leaf scar is approximately 30 mm.

(figure 6.24) were major contributors to Carboniferous coal. The genera are distinguished by differences in the leaf scars on their stems. *Sigillaria* specimens, for instance, have leaf scars on their trunks that look like the seals used by officials in antiquity. Hence the genus name is derived from the Latin *sigillum*, a small sign. *Sigillaria* was one of the two giant arborescent lycophyte types, the other being *Lepidodendron* (figure 6.25).

FIGURE 6.23

Lepidophloios, a Coal Age lycophyte. Photo shows a latex peel of a fossil surface. Length of specimen is 11.3 cm.

FIGURE 6.24

Sigillaria ichthyolepis, another arborescent lycophyte from the Pennsylvanian of Indiana. Length of fossil stem is 15.4 cm.

Lycophytes underwent a pronounced radiation in the Devonian, only to experience great losses in diversity in the later Paleozoic. These extinctions may have been due to the appearance of fungi competent to decay large masses of plant matter. Such fungi might, at least in the beginning, have found the hollow interior of arborescent lycophyte stems more attractive than the solid stems of endarch plants.[13] (Note that some endarchs also have hollow stems.) This fungal rot hypothesis is interesting, but it must be recognized that lycophyte extinctions could have been caused by a number of other fac-

FIGURE 6.25
Geological history of lycophytes. Note their early abundance and the subsequent extinction of most groups by the end of the Permian. *Lepidodendron* and *Sigillaria* are the arborescent forms.

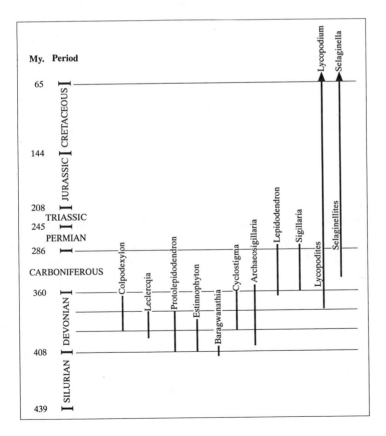

tors, given that some of the lycophyte extinctions coincide with the major late Paleozoic episode of glaciation. Furthermore, recall that small clubmoss lycophytes are still with us, and earlier we hypothesized that the continuing success of *Lycopodium cernuum* might in fact be due to the presence of fungi within its stem.

The latter half of the Devonian also saw the appearance of the sphenophytes, thought to be distant relatives of trimerophyte endarch plants. The sphenophytes, or horsetails, possess conspicuously jointed stems, spores with wind-catching elaters (figure 6.26), and whorled leaves radiating outward from stem joints, as in the genus *Annularia* (figure 6.27). The stem is hollow throughout, with pith surrounding the central air space and discrete, radially arranged vascular bundles producing wedges of secondary xylem. Sphenophyte genera such as *Sphenophyllum*, *Calamites* (figure 6.28), and *Equisetites* became important in the Coal Age. *Sphenophyllum* has a very distinctive triangular primary stele formed exclusively of tracheids. The xylem cells are at the apices of each triangle (figure 6.29). This wood structure is unlike that of any other plant group. Sphenophytes later developed intriguing forms, like the Carboniferous *Amyelon iowense*, which grew stilt roots or prop roots like today's mangroves. The roots of *Amyelon* had an unusual tetrarch (four-sided) stele.[14]

Microspores and Megaspores   The *Calamites* sphenophytes (an endarch group) and *Lepidodendron* lycophytes (an exarch group) show parallel development (convergent evolution) of their reproductive structures during the late Paleozoic. Both groups are thought to have independently developed heterospory, that is, the divergence of spores into two types, microspore and megaspore. Both groups also show telescoping of their fertile parts within the protected female spore-forming organs (megasporangia).

Heterosporous plants produce two sizes of spores (figure

6.30), divergent in shape and size. The larger are the mega-spores, which develop into female gametophyte plants. The microspores grow into male gametophytes. Heterospory is a required precursor stage for the development of reproduction by seeds in plants.[15] Also, heterospory probably allowed plants to colonize drier habitats. For instance, if a plant has two different spore types, the spores can have different sizes and functions, and this reproductive division of labor can have big advantages for a plant species in a dry environment. The small spore can be adapted for desiccation resistance and for long-distance transport. The larger spore will not travel as far, and in fact might be expected to stay close to and perhaps even in intimate contact with the parent sporophyte (this is part of the great inversion of plant development—discussed later—that greatly helped to expand Hypersea). Expansion of vascular plants into cooler and drier environments, aided by hetero-spory, was rapid in the Devonian, and was accompanied by evolution of leaves, roots, and secondary tissues such as wood and bark. And of course, with the development of seeds, it was possible for plants to effectively dispense with the free-growing gametophyte stage, the life stage that requires moist environ-ments.

Because the female and male components of plants today (ovules and pollen grains) show the same size distinctions as sperm and egg, paleobotanists seem predisposed to assume that the large and small propagules of fossil species follow the same pattern. However, there is little evidence as to the type of gametophyte produced by these large and small fossil spores and almost no information regarding their sex.[16] It has been

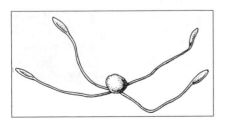

FIGURE 6.26
Spore of modern horsetail (*Equisetum*). The elaters help the spore to disperse by wind. Width of spore, from ends of ela-ters, is 250 microns.

observed that in some living plants, the female-forming spores (megaspores) are actually smaller than the male (microspores). Unexamined assumptions can lead to critical errors when interpreting fossils.

Both microspores and megaspores have been known for some time as fossils in coal balls. Coal balls are calcium carbonate (limestone) nodules formed in coal. These balls are prized by paleontologists because they often contain ancient plant and fungal matter preserved in exquisite detail. Fredda Doris Reed (1894–1988), a professor of plant science at our own Mount Holyoke College from 1928 to 1960, made a paleobotanical career of studying coal balls. In 1926, from an Illinois coal ball, she described well-preserved microspores, as well as a three-dimensionally intact megasporangium showing preservation of three of the four megaspores. In a tribute to Professor Reed on the occasion of her retirement in 1960, Betty F. Thomson (Mount Holyoke Class of 1935) had this to say about coal balls: [17]

FIGURE 6.27 (Left)

Foliage of the Pennsylvanian sphenophyte *Annularia radiata*. All sphenophytes (horsetails) have leaves that radiate from stem joints. Width of shale nodule is 5.2 cm.

FIGURE 6.28 (Right)

Calamites sp., a fossil of the interior of a sphenophyte stem from Pennsylvanian strata of Scranton, Pennsylvania. The fossil is a pith cast, formed when muddy sediment filled the hollow space in the core of a calamitid stem. Such fossils are common in the shaly beds immediately above Carboniferous coal seams. Specimen length is 16 cm.

These are stony lumps found in coal beds that contain agglomerations of fossil plants, usually in fragments but often with their internal structure beautifully preserved. The detective work needed to reconstruct the details of a whole plant from such bits and pieces relies heavily on meticulously accurate observation. And meticulous accuracy is a thing above all others that her students have learned from Miss Reed—both the practice of it and the delight in it, whether in doing their own work or in observing that of others.

Careful observation[18] of fossil spores of the late Devonian plant *Barinophyton citrulliforme* indicates that viable spores of two different sizes, presumed to represent two different sexes, occurred within the same sporangium. The usual situation in extant plants has microspores and megaspores forming in different sporangia. Paleozoic plants such as *Barinophyton* formed spores of different sexes in the same sporangium. Only later were the sexes evolutionarily segregated into different sporangia.

The Devonian fossil *Cystosporites devonicus* provides evidence that the sexes were being segregated into different sporangia. Looking, when magnified, something like a dehydrated zucchini (figures 6.31–6.33), this fossil of a megaspore tetrad has three abortive spores clearly visible at the tip of the megaspore.[19] The *Cystosporites* megaspore is comparable to the spores formed by plants known to segregate spore sexes, and

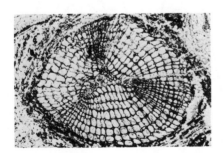

FIGURE 6.29
Sphenophyllum stele. Note the triradial symmetry of the central vascular bundle formed by the characteristic triangle of primary wood. No living plant has this type of stem anatomy. Greatest width of stele is 4 mm.

thus can be interpreted as evidence that the *Cystosporites devonicus* plant had female spores in one type of sporangium (probably the megasporangium) and male spores in the other type.

Strictly defined, a seed is a fertilized megasporangium enclosed within some type of outer covering or integument. There is a tendency in the evolution of plants, seen in the convergent evolution of several groups, toward protection (by integument) of the female megasporangium—a key step in the evolutionary development of seeds. Seeds first appear in the late Devonian, and shortly we will have more to say about seed evolution and its significance for Hypersea.

Roots and Shoots   Turning from the aerial structures of stems, leaves, and spores, what does one find in the fossil record about below-ground structural development? Among the earliest fossil rootlike structures are fine traces of rhizomes from the Silurian Bloomsburg Formation of central Pennsylva-

FIGURE 6.30

Plant propagules leading to the evolution of the seed. A. Sporangium of a homosporous plant with just one size of spore. B. Microsporangium (with microspores) and megasporangium (bearing megaspores) from a heterosporous plant. C. An early seed. D. female gametophyte; E. integument; F. megasporangium; G. megaspore.

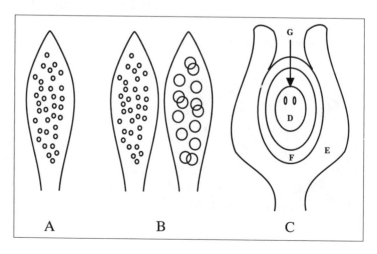

nia.[20] These traces, or impressions, occur in an ancient claystone soil. But massive plant rooting structures are first seen in late Devonian rocks. Meter-size root balls were excavated from Devonian strata during road and dam construction and are now exhibited as a "petrified forest" along Route 990V in Gilboa, New York. These structures, assigned to the genus *Eospermatopteris*, resemble the root clumps of modern palm trees. The centimeter-wide rootlets that form the root ball are tightly packed and parallel, like the pipes in a pipe organ.

By the Coal Age, when Hypersea was certainly well established, arborescent lycophytes had developed the *Stigmaria* type of rooting structure (figure 6.34). *Stigmaria* fossils are sometimes preserved as intact root systems attached to the lower end of an erosion-truncated trunk. In 1851 a *Stigmaria* fossil was offered to the British public as a "fossil serpent." The fossil, "embedded in a coal vein," was advertised by poster as being one of the "wonders of nature" in a "perfect state" of preservation, "eight feet three inches long and seven inches across," and alive "some thousands [sic] of years since." Gentlemen and

FIGURE 6.31 (Left)

Cystosporites devonicus. This genus of megaspore tetrads had one megaspore and three abortive spores. Scanning electron photomicrograph view of entire specimen. Scale bar = 1 mm.

FIGURE 6.32 (Right)

Cystosporites devonicus showing the three abortive spores at the megaspore apex. Scale bar = 200 microns.

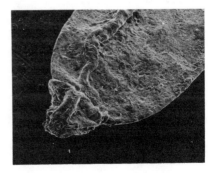

ladies were charged four times the admission fee expected of the "working classes" to view the wonder.[21] The supposed scales on the putative serpent were probably regularly spaced rootlet scars (figure 6.34), and the "natural coil or twist in the trunk" is not unusual in ancient and modern roots, so no doubt the fossil belonged to a plant. Today, *Stigmaria* roots are easily identified in permineralized peats because of hollow structures, which were probably part of an air-conducting system within the roots. Without oxygen in soil (as is often the case in a peat), plants have difficulty utilizing available nutrients. Special aerenchymal tissue in roots can bring oxygen down from above, enabling roots to respire even in water-logged, anoxic soils. The cordgrass (*Spartina alterniflora*) of New England salt marshes is an example of a modern plant with these aerenchymal "lungs." The American Lotus (*Nelumbo lutea*) is another example; a cross section of a lotus stem showing the air passages[22] is shown in figure 6.35.

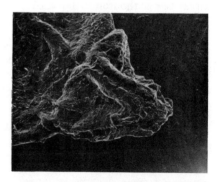

FIGURE 6.34

A piece of fossil root of *Stigmaria ficoides* from Pennsylvanian of McDonough County, Illinois. These roots are often found attached to a trunk. Length of specimen is 23.3 cm.

FIGURE 6.33

Cystosporites devonicus. Close-up view of specimen in figures 6.31 and 6.32. Scale bar = 100 microns.

The Coal Age also witnessed the appearance of the cordaites, endarch plants that grew to tall forest trees over thirty meters high (figure 6.36). They were related to (if not directly ancestral to) modern conifers and were among the first plants to form dense wood. Their leaves were long, strap-shaped, and fibrous; stems had large piths; and secondary steles (branches off the main vascular bundle) showed unusual anastomoses.

Cordaites are notable because they were one of the earliest plants to form seeds provisioned with abundant nutrient reserves. The seed was in the form of a megasporangium that was protected by layers of stony or fleshy integument. The fleshy outer seed coat may have been attractive to animals of the time—if so, it would have been the first fruit. Carboniferous deposits of cordaites indicate that many were mangrovelike in form and in habitat; these are inferred to have grown in salt water. These cordaites had aerial roots like the modern mangrove *Rhizophora*. In Iowa, cordaites coal alternates with seed fern and fern-derived coal of freshwater origin, leading to the interpretation of the Cordaitaleans as being the dominant flora during times of transgression (landward intrusion of the shoreline), while the ferns and seed ferns (discussed below) dominate during times of regression when fresh water returns. In his 1990 book *The Terrestrial Invasion*, Colin Little asserts that the rock strata containing these ancient mangrovelike swamps are likely to be productive sites to search for the origins of early terrestrial vertebrates, since this type of transitional environ-

FIGURE 6.35
Stem cross section of the American Lotus, showing the aerenchyma, or air passages, which bring oxygen to the roots. Similar structures are found in *Stigmaria* fossils (figure 6.34). Width of specimen is 11 mm.

ment might have been used by our ancestors making the leap to terrestrial life. Other cordaites are found away from the swamps, on better drained, sandy soils in upland habitats.[23]

Spores, Seeds, and Stems   The late Devonian and early Carboniferous also gave rise to a variety of winglike appendages on the propagules of spore- and seed-bearing plants. The first evidence for adaptations that facilitate wind dispersal of propagules comes from fossils of small seeds with winged side flanges from the upper Devonian strata of central Germany.[24] This particular Devonian seed is less than 5 mm in length and resembles a tiny snowshoe.

Spore-bearing plants developed their own version of this dispersal strategy. The spore-carrying structure *Tetrapterites visensis* from the Forest of Dean, Wales, is a bizarrely ornamented tetrahedron (figure 6.37). Its odd shape would have

FIGURE 6.36

Reconstruction of cordaites. Height is approximately 10 m.

assisted the transport of the spore through the air or water. This wind-catching spore carrier, called a skiadion, had a cup-shaped enclosure to hold the spore during transport.[25]

This wind-catching skiadion is functionally equivalent to the four straplike appendages called elaters, which add buoyancy to the spores of the modern sphenophyte *Equisetum* (figure 6.26). Wall ornamentation, as in these two examples, reduces the effective density of the spore by promoting a thicker boundary layer of air. This effect increases the buoyancy of the spore in air the same way that the dimples on a golf ball or the fluff on a dandelion seed assist flight.

Spore ornamentation can also enhance animal transport or possibly serve as a deterrent to small spore-feeding herbivores. Another early Carboniferous megaspore, the 2-millimeter-long *Triletes pannosus*, has hooklike spines that project 150 microns (figure 6.38). The spines certainly look capable of grasping onto something, and they probably functioned in just this way, enhancing dispersal of the spores by hitchhiking on passing animals.[26]

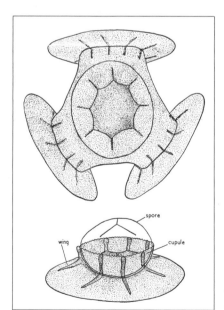

FIGURE 6.37

Tetrapterites visensis. This early Carboniferous spore-bearing structure is called a skiadion. Its unusual shape would have made it easily transportable by wind or water. Greatest dimension of fossil is approximately 210 microns. From reference 25.

Psaronius was a distinctive Coal Age fern of the order Mara-
ttiales. Marattialean ferns had a linear arrangement of sporangia into padlike structures (called synangia) on the undersides
of their leaves. These ferns formed a trunk more than 60 cm
in diameter near the base, tapering gradually to 15 cm in
diameter near the top. The swollen basal trunk of Psaronius,
however, was actually a thick mantle of root tissue surrounding
a rather slender stem. The vascular system of the stem was
anatomically quite complex. It was composed of imbricate bundles of steles. Each stele was composed of a core of xylem
surrounded by a phloem sheath, and the whole bundle was set
in a matrix of pith. This kind of stem (called a dictyostelous
stem) evolved a number of times in the history of land plants.
Its apparent attribute is that it may be a relatively easy way to
develop a thick trunk that will allow the plant to grow higher

FIGURE 6.38

Triletes pannosus, a spore with hooked spines that suggest animal transport
of spores in the early Carboniferous. A. General view of specimen. B, C.
Spines, short hairs, and papillae in profile. D. Part of the surface of the
spore wall, showing a single buttressed spine and papillae. E. Part of the
wall in surface view. From reference 26.

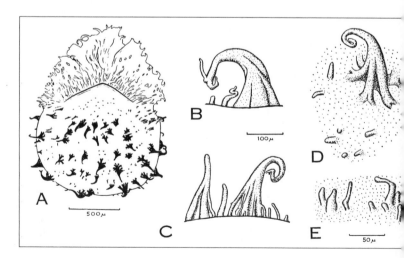

in order to capture enough light in the midst of its leafy competitors.

Basal trunk fossils of *Psaronius* from the Coal Age have been found which, upon sectioning and polishing, show astonishingly brilliant hues. German jewelers cut these fossils and make them into brooches, sold as *Staar-Steine*. This name translates as "Starling Stones," in reference to the fact that the petrified rootlets resemble, in cross section, the speckled winter plumage of the European starling. The native Hawaiian tree fern *Cibotium chamissoi*, called *hapuu* by the Hawaiians, has a similar but independently (convergently) evolved arrangement of roots to form its trunk. The stem in the Hawaiian species is surrounded by a dense mass of aerial roots, which not only absorb nourishment for the fern but also aid in supporting it in an upright position and perhaps in protecting the core stem from injury.

The mild global temperatures prevailing during the Devonian and early Carboniferous, combined with favorable precipitation patterns in what are now North America and Europe, were responsible for a largely cosmopolitan flora that produced tremendous coal reserves. The Mississippian coal forests were dominated by lycophytes, but crustal disturbances and related mid-Pennsylvanian climatic fluctuations allowed different groups to dominate at different times.

One of these new groups was the pteridosperms or seed ferns, another group of endarchs that first appeared in the early Devonian. This distinctive group of plants lived until the Jurassic; shortly afterward, modern angiosperms began to dominate the land flora. Seed ferns seem to have been the first plants to make use of pollen, the ultraminiaturized propagule of the male gametophyte.

The foliage of seed fern species is in many cases unknown. However, these plants are thought to have had fernlike leaves. *Neuropteris* (figures 6.39–6.41) is the name given to a common fossil foliage of the late Paleozoic thought to belong to pteridosperms. Immature *Neuropteris* foliage has been found in direct association with fossil pteridosperm stems. *Neuropteris* foliage is thus indicative of the presence of pteridosperms.[27]

Seed ferns reached perhaps five meters in height. The lyginopterid pteridosperms had a single stele in stems and an enclosed seed. Plates of thick-walled cells occurred in the stem cortex. The medullosan pteridosperms (such as *Neuropteris*) had several steles in a given stem, much like the marattialean ferns mentioned earlier. This multistem morphology must be a favorite focal point for convergent evolution, as we see it appearing again and again. Medullosans had fused sporangia and a solid, stony integument surrounding the seeds.

Seed ferns such as the medullosans and the lyginopterids are credited with the invention of pollen. Pollen is in essence an ultraminiaturized and abbreviated male gametophyte stage, packaged for long distance transport. There is really only one cell in this miniature gametophyte. The cell has, nevertheless, multiple nuclei of two types. One, the tube nucleus, coordinates growth of the pollen tube. Nuclei of the other type, the generative nuclei, drift down the pollen tube to complete fertilization of the seed.

In addition to being the first plants to utilize pollen, pteridosperms were noteworthy for bearing the largest gymnosperm seeds of all time (several centimeters in length). The medullosan seed *Pachytesta* had a thick integument consisting of a fleshy outer layer and a hard inner layer. It looked like a large lollipop stuck on the tip of a branch.

FIGURE 6.39

Neuropteris, the foliage of the pteridosperms or seed ferns. Height of specimen is 3 cm.

Seed ferns give both direct and indirect evidence of plant-animal interactions in the Paleozoic. Some *Neuropteris* leaves from the famous Mazon Creek locality of Illinois (late Carboniferous) have clearly been bitten or chewed. A nick on the edge of the leaf in figure 6.40 may indicate insect damage of this sort. There is also some evidence for leaf-mining in Coal Age fossil foliage. Medullosan seed fern pollen is often so large that wind pollination was unlikely; thus, arthropods were probably the agents of pollination (the pollen probably stuck to the early pollinators). Miniaturization of pollen, like miniaturization of computers, was a sequential process.

The occurrence of both *Neuropteris* and *Cardiopteridium* (which had neuropteran foliage with heart- rather than blade-shaped leaflets) in North America gives evidence that early Coal Age floras in the northern hemisphere were nearly cosmopolitan. But by the late Pennsylvanian, tectonic isolation had produced four distinct floral provinces (figure 6.42). The Euroamerican province continued to support lush swampy coal forests. The Angara (Siberian) province became more temperate, and lycophytes there gave way to cordaites. The Cathaysian (Oriental) province remained tropical to semitropical and de-

FIGURE 6.40 (Left)
Neuropteris decipiens from the Pennsylvanian Mazon Creek locality of Illinois. The small notch in the frond at the upper left may indicate herbivore damage. Length of fossil is 11.4 cm.

FIGURE 6.41 (Right)
Neuropteris loschii, Pennsylvanian of Clinton County, Missouri. Greatest width of rock specimen in view is 11.6 cm.

veloped distinctive floral forms such as the earliest gingko ancestors.

The Gondwana province of the southern Gondwana continental landmass had a temperate climate. *Glossopteris* occurs as a fossil in all continents that were once part of Gondwana, but its antecedents are unknown. The dominant plants of the glossopterid flora, exemplified by the genus *Glossopteris*, were interesting forms with lance-shaped leaves and reproductive structures on stalks that projected from the central leaf vein.

FIGURE 6.42.
Pennsylvanian Floral Provinces. World map (modern continental positions, nonstandard Mercator projection) showing the Euramerican, Gondwanan, Angaran, and Cathaysian floral provinces of the late Paleozoic.

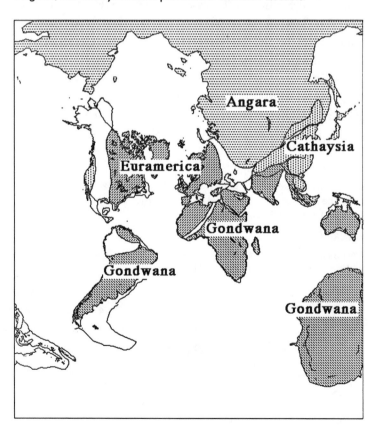

Large wedge-shaped sections of stems devoid of cells in the glossopterid genus *Vertebraria* are thought to be parts of an air-conducting system.

Profound changes in climate such as continental glaciation took place in the late Carboniferous and foreshadowed the widespread aridity of the Permian. These climatic changes have been linked to the late Paleozoic tectonic assembly of the supercontinent Pangea. This supercontinent formed by the continental drift and collision of nearly all of Earth's continents to form a huge land mass. Coal Age swamp plants were ill-adapted to the increasing aridity of the supercontinent-dominated climate and were replaced by more hardy vegetation. Sub-floras conditioned to a drier climate appeared during the early Permian in what is now the southwestern United States. Permian plants lived in less lush and less crowded conditions than had their Carboniferous forebears, and they developed large stiff leaves. As the Paleozoic ended, *Glossopteris*, early conifers, and xeric (dry-adapted) descendants of seed ferns replaced the drought-intolerant swamp plants. By the close of the Permian, arborescent lycophytes, lyginopterids, medullosans, and cordaites had disappeared from the scene. The new xeric forms of seed ferns and calamites, as well as conifers, cycads, and cycadophytes, became the globally dominant terrestrial flora.

In 1956 Hilda Stoneley discovered an ancient plant that might qualify as the earliest succulent, on account of its extraordinarily thick, desiccation-resistant leaf cuticle. This fossil, which she named *Hiltonia*, occurs in Permian desert sandstone deposits. In *Hiltonia*, we see renewed emphasis on the part of a plant to preserve its moisture content, as part and parcel of its growth in a dry and occasionally hot habitat.[28]

In summary, the first vascular plants are likely to have appeared as long ago as the early Silurian. By the end of the Silurian, rhizoids of vascular plants had appeared, and fossil evidence suggests that these rhizoids were likely to have been associated with fungi. By the early Devonian (Siegenian Age), microphylls had appeared in the exarch lineage of vascular plants. A few million years later (the Emsian Age of the early Devonian), larger leaves or leaflike organs (megaphylls) enter the fossil record. Secondary xylem made its debut in the middle Devonian (Eifelian Age), and arborescence or tree-shape evolved during or close to the Givetian Age. Deep roots or rhizophores appeared in the late Devonian (Frasnian Age). The late Devonian Fammenian Age marked the first arrival of both leaf abscission (ability to shed leaves) and the first seeds. Seeds have a greater survival potential than do tiny spores, and the development of seeds permitted the widespread colonization of continental interior habitats by vascular plants such as cordaites. Leaf abscission also is a valuable adaptation for xeric conditions, and it promoted the accumulation of plant litter. The ecological significance of plant litter for the evolution of land animals is the focus of the "litter box" hypothesis, treated in chapter 7.

The evolution of vascular plants through the Paleozoic is best summarized by its major trends: loss of many early character combinations, propagule miniaturization, development of secondary tissues such as wood or fused multiple stems for support, the move toward airborne or animal-carried propagules, and admission of air into plant tissues via specially adapted gas pockets. All these traits evolved independently in unrelated groups of plants, suggesting that they represent viable responses to some very strong evolutionary selection pressures.

Plant Development Turned Inside Out

Popcorn is one of our favorite snacks and flavor isn't the only reason. It illustrates angiosperm reproduction and provides a metaphor for evolution of flowering plants.

The popcorn kernel, like the seeds of most familiar angiosperm crop plants, is primarily a starchy substance called endosperm. From the plant's point of view, the purpose of endosperm is to nourish the embryo—its fraternal twin. Precursors of the embryo and the endosperm are fertilized at the same time, which makes provisioning the embryo much easier on the parent plant than is the case in non-angiosperm methods of fertilization.

Non-angiosperms store all the embryo's nutrition within the embryo sacs before the eggs are even fertilized. This explains in part why pine nuts, a gourmet delicacy, are so much more expensive than, say, almonds. In the gymnosperm ("naked seed") piñon pine, it takes a relatively long time for the embryo sac to fill with nutrients. Also, the pine's way of producing offspring is the biological equivalent of putting down a month's rent in advance, since the embryo hasn't even been fertilized yet. In flowering plants, the embryo is paired with a fast-growing endosperm "helper" that swells with nutrients to be stored for the embryo.

Endosperm storage is faster and more reliable than egg sac storage for two reasons. First, the endosperm contains a half copy of the paternal genes and a full copy of the maternal genes, so it is less likely to be defective because of a faulty copy of one gene, a possible point of failure for the maternally derived embryo sac cells. Second, because it has multiple copies of genes, from both parents, the endosperm can amass greater stores of food faster. The three thirds of the genetic complement work in concert to develop the food store.

Angiosperms, then, have a supercharged embryo that contributes to their evolutionary success. This new plant embryo type first appeared in the Mesozoic about 110 million years ago.[29] By the late Mesozoic angiosperms had displaced many of the plant groups that previously had been so successful.

Angiosperms were the first to realize the full potential of the plant embryo. But others had prepared the way. The development of the first plant embryos, and thus the appearance of embryophytes or true plants, occurred in the Paleozoic. Interestingly, the development of embryos also involved a nutritional change in a plant propagule. This change is key to understanding plant evolution, for it involved the development of an intimate relationship between the haploid (gametophyte) and diploid (sporophyte) halves of the life cycle.

In a typical fern, for example, the sporophyte plant is far bigger and longer-lived than the gametophyte plant, but for most of their respective lives they each live and photosynthesize independently and sequentially. Despite this independence, the fern gametophyte provides a nurturing environment for the embryonic sporophyte. Ever since the first embryos, one part of the plant life cycle has clearly nurtured, fed, and protected the other part of the life cycle.

A nutritional relationship between the haploid and diploid phases of the life cycle, similar to what now occurs in embryophytes, apparently first evolved in the Paleozoic in an organism similar to the modern freshwater green alga, *Coleochaete* (figure 6.43). In *Coleochaete*, the egg is retained within gametophytic tissue. Recent studies support Frederick O. Bower's hypotheses of 1908 concerning the alternation of generations and the origin of sporophytes in land plants.

Bower's hypothesis states, in essence, that the sporophyte generation (the conspicuous vegetative stage in familiar vascular plants) developed *de novo* from a haploid alga that lacked a diploid sporophyte generation but instead had merely a diploid zygote (a cell or embryo formed by the fusion of two gametes

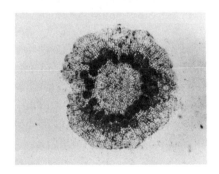

FIGURE 6.43

Coleochaete, an aquatic charophyte green alga. Whole thallus showing concentric rings of zygotes; youngest zygotes peripheral to more mature, darker ones. All land plants evolved from a Paleozoic organism very similar to this. Except for the zygotes, all the cells seen in this photograph are haploid. Width (entire organism) is 1 mm.

or sex cells). Before the evolution of embryos, this zygote would have immediately undergone meiosis (to relieve the diploid condition) and produced spores, the propagules of the next haploid generation. Growth of such a spore is analogous to having a hypothetical haploid human generation developing from an isolated sperm or egg cell. The implication here is that different plant groups developed the sporophyte generation independently, by convergent evolution.

An older, competing hypothesis dating back to 1874 held that the algal ancestor of embryophytes already had had alternation of two generations for a long time (see "Plants Defined" in chapter 4) and was thus diplobiontic, as opposed to haplobiontic. Haplobiontic organisms, such as humans, have the gametes or sex cells as the only haploid cells; diplobiontic organisms develop those haploid cells into a multicellular life stage. The diplobiontic hypothesis of 1874 is less favored now because it fails to explain how the sporophytes and gametophytes, which in modern diplobiontic green algae have no long-term physical connection, could have evolved the intimate physical connection, in both nutritional and developmental respects, shared by the haploid and diploid components of all embryophytes.[30]

The 1874 hypothesis requires a preexisting sporophyte generation to attach onto and to even become parasitic upon its own counterpart haploid generation—not a totally implausible scenario, just one that seems to us to be less likely and less supported by the evidence. The sporophyte generation probably forms by delayed meiosis in the zygote, coupled with retention of the zygote within the haploid tissues of the gametophyte.

With Bower's hypothesis, we see a fascinating exchange between haploid and diploid life cycles, which play incubator and host to one another within their own tissues. The sporophyte generation first appeared as an added generation that came into existence as a result of delayed zygotic meiosis—sort of a delayed plant puberty. In other words, what might otherwise have become the new haploid cells of the next generation by chromosome reduction instead retained its diploid character and thus added, à la Bower, a new generation to the life cycle. In previous, ancestral forms, the diploid cells were short-lived intermediaries en route to the creation of haploid propagules in spore form. After the switch to alternation of generations, what was previously a few diploid cells developing into spores ("puberty") began to delay the final step. The final step of spore genesis still eventually occurred, but not until after the diploid cells had grown and developed into a new sporophyte generation, in essence an overgrown zygote.

Enclosed by the gametophyte from the start, the newly added sporophyte generation with a double dose of chromosomes was enveloped by nurturing haploid tissue. In *Coleochaete*, a modern freshwater chlorophyte, a layer of haploid cells covers the surface of the mature zygote (figure 6.43). These cells form ingrowths that penetrate the zygote to provide nutrition. The similarities with fungi are intriguing; the vegetative haploid cells of *Coleochaete pulvinata* look like branched, septate fungal hyphae. The protected diploid zygote in *Coleochaete* gives the aquatic alga advantages because many more spores can be produced from a single fertilization event than would be the case if the zygote hurried straight to meiosis and the formation of one of those four spore tetrads so common in the fossil record.[31] Under Brower's hypothesis, we suppose that, from the point of view of the gametophyte, the sporophyte generation is like a giant multicellular spore factory.

We hypothesize that the additional complement of spores made possible by this balancing of the two stages of the life cycle proved biologically "worth" the investment made by the gametophyte to nurture the now parasitic zygote. The hypothesis remains to be tested, but if it is correct, the sporophyte generation (diploid cells) came to develop inside (and be protected by) the gametophyte generation (haploid cells) precisely because the arrangement ultimately benefited both generations. Later, of course, some sporophytes evolved to be independent of the gametophyte generation and to be free-living plants.

It is actually quite surprising that the sporophyte generation, a rudimentary generation in the earliest stages of plant evolution, "took over" in plants that evolved later. In subsequent plant evolution the roles of nurturer and nurtured played by the alternating generations actually reversed. In seed plants, the gametophyte generation is now completely (with the exception of the pollen) enclosed in the dominant yet protective sporophyte generation.

Plant life cycles thus were literally turned inside out. Why did this reversal occur? And how? The first question is fairly easily (and uncontroversially) answered, the second is less straightforward but has a direct bearing on Hypersea theory.

Ecological Causes of Life Cycle Changes

Early embryophytes had both a vegetative sporophyte (descended from a *Coleochaete*-like zygote) and a vegetative gametophyte. As competition for light increased in early land ecosystems, the sporophyte generation was favored because of stringent limits to how high a gametophytic plant could grow without stranding its gamete offspring. A gametophyte plant requires moist conditions in order for swimming sperm to reach the archegonium (egg or ovum chamber), fuse with ova, and form the zygotes. Because wind speed and attendant desiccating effects increase dramatically with the height of a plant, "high and dry" is not a happy condition for a gametophyte plant trying to reproduce. Thus, sporophyte plants, able to fabricate and release copious quantities of desiccation-resistant spores, gained the upper hand in Paleozoic terrestrial communities.[32]

This same "high and dry" effect explains the relative rarity of modern free-living gametophytes in a kingdom of sessile organisms. Most people are familiar with the vegetative fern sporophyte, but who among us recognizes a fern gametophyte or the gametophyte of a bryophyte plant? In 1848 a leading

authority on clubmosses wrote that, as he had never seen clubmosses (*Lycopodium*; see figure 6.16) produce seed, he suspected that all the female plants had been destroyed in the
Biblical Flood. Apparently the botanist mistook clubmosses
and their asexual spores to be flowering plants bearing highly
specialized "unisex" flowers that produce pollen grains but no
ovules that could mature into seeds.

The mystery of *Lycopodium*'s life history was solved in Java
in 1886 when gametophytes of *L. cernuum* were finally located.
In 1922 the greatest number up to that time of gametophytes
of any living lycophyte species were discovered when the sexual
stages of *L. cernuum* appeared by the thousands at the Kilauea
crater in Hawaii in the volcanically heated crevices of the
Sulfur Bank and its vicinity. This population formed something of a volcanic vent biota of the land. In 1930 Otto
Degener[33] wrote:

> The rarity of the gametophyte under normal conditions and its
> phenomenal occurrence in fumaroles is due to the fact that the
> ideal environment for the growth of the one generation is
> harmful to that of the other. Such an explanation may, in part,
> account for the extinction of most of the clubmoss relatives
> which flourished during the Carboniferous Period and largely
> built up our coal deposits with their branches, leaves and
> inflammable spores.

Degener's suggestion for the extinction of gametophyte-domi-
nant Paleozoic plants accords with their need for high humidity
for gamete dispersal. It implies that if hot, moist, and possibly
carbon dioxide—rich conditions were to return to Earth, gametophytes might once again become abundant. It also highlights the importance of the protective seed coat in plants that
form seeds. In clubmosses, if the conditions for growth are not
ideal for the embryonic sporophyte (embedded within the tissue
of a female gametophyte) when it falls to the ground, it will

soon perish. Both generations tumble together out of the cone of a clubmoss (figure 6.16) and together seek their fate without the security of a protective coating as is the case for seed plant embryos.

Lycopodium cernuum gametophytes do, in fact, occur in abundance in areas other than those adjacent to volcanically heated crevices. In peninsular Malaysia they are found in great numbers on moist shaded banks. The tissue contact between the gametophyte and sporophyte generations in *L. cernuum* occurs across tissue called the placenta; the wall ingrowths of the two generations in this zone are so closely juxtaposed that they appear almost to merge.[34]

Recall our earlier mention of fungi inside of *L. cernuum*, and our suggestion that these fungi might have helped the species survive. Both the gametophyte and sporophyte cells of *L. cernuum* are invaded by fungi. Figure 6.44 shows fungal hyphae in the tissue of an *L. cernuum* gametophyte. Two large chloroplasts are visible in the figure. Figure 6.45 shows coils of hyphae in gametophyte epidermal cells. The fungi have been surrounded by host cell wall material, a sign that the hyphae have been inside the plant cell for some time. Figure 6.46 is an oblique section through large hyphae in the gametophyte—note the large lipid (oil) droplets. Figure 6.47 shows thin-walled fungal hyphae within the *L. cernuum* gametophyte. Breakdown of fungal hyphae releases lipids to nourish *L. cernuum*, and evidence (in the form of skinny chloroplasts, as opposed to starch-filled chloroplasts) suggests that photosynthetically produced food is translocated from *L. cernuum* gametophyte cells to the fungus.

In the whisk fern *Psilotum*, considered by botanists to be the most "primitive" living vascular plant, similar lipid-packed fungal hyphae in the cells of its colorless (chlorophyll-lacking) gametophyte are thought to provide the gametophyte with food by releasing the lipid reserves when the fungal hyphae degenerate. The fungi, naturally, must get food from another source— probably in this case, soil organic matter. The whisk fern gametophyte thus acts as a parasite, and after the lipid-bearing fungi degenerate, the plant is often reinfected by fresh fungal hyphae. There must be some benefit to the fungus, or else it wouldn't keep coming back with its loads of lipid!

Recall the problem, mentioned earlier, of paleobotanists wondering how foodstuffs derived from photosynthesis were able to get to the roots of arborescent lycophytes, considering that the plants lacked phloem. We hypothesize here that, as is the case with the whisk fern and with *Lycopodium*, the roots of lycophytes received part or perhaps most of their nutritional needs from infectious fungi. Could such fungi even provide services usually rendered by phloem tissue? The serpentine roots of arborescent lycophytes may have had more of what we call a Hypersea aspect than any paleobotanist has ever expected. Judging from the abundance of fungal remains in coal, the postulated root fungi of lycopods would have had access to more than sufficient food supplies.

We must now return briefly to the topic of evolution of sporophyte dominance. In some lineages the gametophyte came to be entirely retained within the sporophyte generation; that is, the haploid spores developed and matured into gametophytes within the diploid sporophyte tissues. Since spores could no

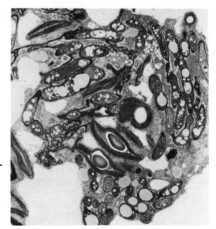

FIGURE 6.44

A cross section through the gametophyte plant of *Lycopodium cernuum* shows how fungal hyphae incorporate themselves into the tissue of the plant. The irregular, sausage-shaped or ovoid structures are sections through the hyphae; the two eye-shaped structures below center are sections through chloroplasts. Width of photomicrograph 16 microns.

longer be the propagule of choice for female gametophytes, the male gametophyte generation had to take on all the travel responsibilities. Pollen is essentially a haploid male gametophyte that never photosynthesizes; its female counterpart never even sees the light of day.

Pollen grains are very resistant to desiccation but can explode and die when they come into contact with relatively pure water. Even rain is dangerous because of the spores' very low internal osmotic water pressure. It is somewhat ironic that the male of the moisture-loving gametophyte generation has become encapsulated in such a way that it can be killed by an overabundance of water. It might seem that sporophyte dominance led to a "sterilization" of the life of the gametophyte generation, in the sense that, unlike the case in *Lycopodium*, neither the male pollen nor the female gametophyte has contact with symbiotic fungi. The tendrils of Hypersea, however, reach even to the domain of the encapsulated gametophyte

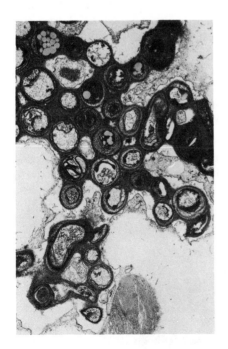

FIGURE 6.45
Coils of hyphae infuse the epidermal cells of a *Lycopodium cernuum* gametophyte. The stacked, circular structures are closely spaced sections through tightly coiled hyphae. Each hypha is cut numerous times, the same as if a tightly coiled spring were to be sliced along its longitudinal axis. The fungal coils have been surrounded by dark, cell wall material of the host—a sign that the hyphae have been inside the plant cell for some time. Width of photomicrograph is 5.3 microns.

generation, as will be shown in the discussion of vigorous tall fescue to follow.

Triumph of the Sporophyte and the Expansion of Hypersea

Encapsulation of the gametophyte stage within the sporophyte thus made terrestrial plants capable of being even more terrestrial—they no longer needed a very moist environment for dispersing gametes. The "why" of encapsulation is rather obvious, but what about the "how"? How did encapsulation of the sporophyte occur?

As already noted, the evolution of delayed zygotic meiosis can lead to a larger zygote and ultimately more spores. Developmental delays like this are a relatively simple modification commonly seen in evolution; the delayed meiosis could confer an immediate reproductive advantage by getting more spores out of each fertilization event at the relatively modest cost to

FIGURE 6.46 (Left)
Oblique section through fungal hyphae living inside the cells of a *Lycopodium* gametophyte. The pale spaces inside the hyphae are lipid droplets. The lipids are released to nourish the plant when the hyphae die. Who is providing nutrients to whom? Width of photomicrograph is 5.2 microns.

FIGURE 6.47 (Right)
A further magnification of the thin-walled hyphae inside a *Lycopodium* gametophyte. These nonseptate hyphae indicate that the symbiotic fungus is probably a zygomycete fungus. Width of photomicrograph is 3.2 microns.

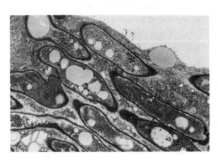 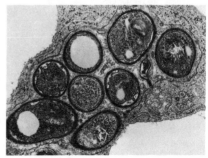

the parent gametophyte of some investment in nutrition to build the sporophyte spore factory. Such an advantage could certainly get the process started in an evolutionary sense, and it may even explain progressive enlargement of the sporophyte (because the larger the size a sporophyte obtains before meiosis, the more spores it can produce). Thus, a large sporophyte can evolve in a gradual fashion, without evolutionary jumps, just by continually improving its reproductive success. We do not mean to imply that this evolution could not have happened abruptly, and indeed large sporophytes may have developed over a very short stretch of geologic time. But the change is one of degree: there was no substantial "jump" in body structure or reorganization.

The late Silurian to early Devonian genera *Parka* and *Pachytheca*, which in one prime locality are found in thin but extensive fossil deposits, resemble the modern freshwater alga *Coleochaete* in structure and habitat. Selective pressures (such as high spore mortality) in Silurian and earlier shoreline habitats may have favored production of large zygotes and greater numbers of spores in these ancient genera. *Parka*, which was a small green shoreline organism ecologically associated with the exarch *Zosterophyllum* (but taxonomically on fundamentally different paths), produced numerous spores. *Parka* spores were formed in large sporangia covered by a cellular layer reminiscent of the cellular layer that coats the zygote of *Coleochaete* today.

The adaptive significance of zygote retention in *Coleochaete* is that it allows the zygotes to remain in shallow water over the winter. Zygotes released by other green algae are dispersed and many settle in deeper water and may be at a disadvantage during spring colonization season: the competing *Coleochaete* zygotes (at anchor during the winter in parental gametophyte tissue) are released directly into the shallow nearshore waters when spring arrives.

Recall that adaptation to local environments can favor a large zygote in an aquatic green algal lineage if the local environment is conducive to the survival of species that produce large numbers of spores. In an evolutionary sense, this seems quite straightforward. On a deeper evolutionary level, however, this "endozygotic event" takes on a much greater significance than the mere enhanced production of spores or even the origin of embryophytes.

The importance of the intimate relationship between the alternate generations is the habitat the one is providing for the other—a habitat that can be regarded as an inner sea, Hypersea. This new habitat was completely unexploited and unexploitable before the endozygotic event that led to large sporophytes. This event cannot even be considered to be an example of symbiogenesis because a creature cannot by definition be in symbiosis with members of its own species, let alone its own parent. We consider such cases to be evidence, rather, of hypermarine interactions between plant generations. The "inner sea" then is eukaryotic body fluid; organisms able to access and colonize this sea launched an evolutionary sequence that led to transformation of the land surface.

Lichens are a rudimentary stage in the genesis of Hypersea. Fungal hyphae trap and enslave green algae and then, incidentally but significantly, the phycobiont-mycobiont complex gains the ability to grow over formerly sterile rock surfaces. Even if one views lichens as a case of controlled parasitism, the lichen can also be thought of as a mycobiont providing a watery environment whose boundaries (shorelines) are stretched and expanded by its associate phycobionts. Indeed, the typical crustose lichen (figure 4.2) resembles a series of wave crests radiating outward from a common center—a living tide that flows out over the surface of its rock or bark substrate.

The triumph of the plant sporophyte is an even more pronounced example of the spread of Hypersea. In the beginning, the interior sporophyte generation (much like the lichen's phycobiont) is surrounded by the vegetative haploid gametophyte (analogous to the haploid mycobiont of a lichen). The analogy holds up; the haploid growing edge of *Coleochaete pulvinata* even looks somewhat like a branched, septate fungal mycelium with chloroplasts added.

There is one notable difference, however, between the way lichens work and the way the sporophyte-gametophyte couplet works: in lichens, nutrients flow mostly from phycobiont to mycobiont; in early-type embryophytes, where the gametophyte is the primary plant, the flow is from gametophyte (outside) to the sporophyte (inside). In modern-type plants, where the sporophyte is dominant, the nutrient dependence is reversed: the sporophyte has broken through the surface of its own, parental "inner sea." In the modern-type plants, we see that the gametophyte has become the enclosed, dependent member of the pair.

This inversion of the gametophyte-sporophyte relationship in the Paleozoic altered the face of the planet and very soon led to a tremendous increase in global biomass—an increase of nearly two orders of magnitude. Like a living wave rising up from the gametophytic inner sea, sporophytes splashed across the earth's surface in the Paleozoic. An initial turning inward, an "autosymbiosis" in a sense, proved to be the real key to development of a conspicuous terrestrial biota. Sporophyte dominance helped free plants from their requirement of moist environments.

Now, getting back to popcorn . . . Popcorn is the seed of a mutant plant that cannot reproduce in the wild because it needs human assistance to husk the ear and free the seeds. The corn kernel is a tiny sporophyte plant directly descended from a Paleozoic green algal zygote. The corn kernel plus its nutritious fraternal endosperm are conceived by and cared for by a gametophyte which in turn is nested within a mature sporophyte. The Paleozoic sporophytes that broke through the bounds of the gametophyte can be regarded as popping kernels—an event that was "heard" around the world.

The tendency of plants to evolve linkages between generations can reach right past the gametophyte, linking sporophyte with sporophyte. A fine example grows in our own herb garden. It is the Egyptian Onion (*Allium cepa,* variety *viviparum*), also called the Walking Onion. Our variety of this food crop of ancient pedigree bears flowers on stalks which are followed not just by seeds, but also by small bulblets or bulbils. Each of these bulbils is an entire miniature onion sporophyte nourished by the parent sporophyte. When the parent stalk ages, it weakens and droops to the ground, allowing the bulbils to root. This action accounts for the name Walking Onion. The bulbils thus act as a precocious type of propagule.

In the particular breed of onion we grow, some of the bulbils actually develop stalks while still attached to the mother plant, and these stalks can also form tiny but viable, second order bulbils without touching the ground. It seems possible that a very robust mother plant could produce third order bulblets. With nutrients still flowing up the main parental stalk, two (and perhaps three) generations of hyper-onions have fully linked inner seas. Bulbils may be a very ancient form of propagation, for they are also formed by the living lycophyte *Lycopodium lucidulum,* a close relative of *Lycopodium cernuum* discussed earlier.

Strange Hydraulics in Devonian Fossils

The discussion in the previous section might erroneously give the impression that the evolution of advanced plants was a linear affair, a relentless march in one direction, with each new organism preparing the way for even more complex forms that were capable of projecting Hypersea into ever more desiccated environments. This next section should demonstrate that not all early Hypersea organisms were on the direct line to chestnut trees.

Spongiophytes, an extinct group, are an interesting type of Devonian land organism with a decidedly low-relief, "primitive" aspect. The spongiophytes are ground creepers. They are characterized by a flattened, branching body or thallus with a thick, resistant, and flexible cuticle. The interior of the thallus is thought to have had a spongy construction of loosely packed living cells. (It is this feature that sets spongiophytes apart from the filamentous nematophytes, soon to be discussed.) Figure 6.48 shows a reconstruction of *Spongiophyton minitissumum* from the latter part of the early Devonian of New Brunswick and Quebec. The branching thallus has both pores and bumpy protuberances on its upper surface.

Spongiophytes had a resistant cuticle, which sets them apart from their aquatic ancestors; a cuticle this thick would probably have inhibited photosynthesis under water. The spongiophyte *Orestovia,* from the Voronezh region of Russia, had pores in its outer cuticle. On the basis of putative guard cells associated with the pores, these pores were formerly interpreted as stomates. *Orestovia* fossils also have apparent conducting cells with spiral and annular thickenings; these resemble the conducting cells of vascular plants.[35]

New evidence, however, from well-preserved specimens from the south shore of the Gaspé Peninsula of eastern Canada suggests that *Spongiophyton* was a lichen.[36] According to paleobotanist William E. Stein, *Spongiophyton* had a thick cortex or tissue layer forming the upper cuticle of its thallus, and a thinner cortex forming the underside of the thallus. This com-

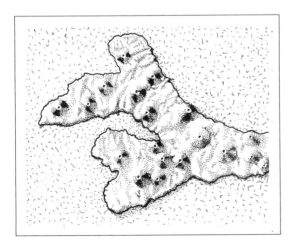

FIGURE 6.48

Spongiophyton minitissumum, a prostrate Devonian spongiophyte. This fossil may represent an early lichen. Pores on the upper surface would have allowed gas exchange; they bear microscopic hyphae. Width is approximately 2.5 cm.

bination of cortical tissue (cortex) is seen in some modern lichens. Stein has also found an apparent holdfast structure on *Spongiophyton* fossils. But the most telling evidence is the reticulate meshwork of hyphae on the inside of the fossil.

Stein interprets this hyphal network as the mycobiont of a lichen. Hyphae appear to fill both the pores and perforate tubes within *Spongiophyton*. Further evidence based on study of the interior of the thallus in *Spongiophyton minitissumum* suggests the presence of the phycobiont layer typical of modern lichens. Stein has found blob-shaped spheroidal cells on the interior of *Spongiophyton* surrounded by fungal hyphae. We agree that these blob cells could be the phycobiont. Stein notes that spongiophytes have cup-shaped structures on their upper thallus surfaces that look very much like structures formed by extant, lichen-forming ascomycete fungi.

In light of this new interpretation of spongiophytes, it appears that lichens may have been much more abundant in the Devonian than previously thought. At his lectures, Stein shows a slide of a painting of a Devonian landscape filled with early vascular plants. He then asks the audience, "What is wrong with this picture?" The answer, in his view, is that the paleolandscape reconstruction fails to show any lichens.

Some specimens of the spongiophyte *Orestovia* are thought to have been parasitized by an ascomycete fungus. These fungi, possibly members of a group called the hemisphaeriacean fungi, apparently formed fruiting bodies (ascocarps) on the under surface of *Orestovia*. The relationship of these hemisphaeriacean fungi to the presumed mycobiont of the spongiophyte is unclear. Perhaps they are one and the same fungus. Alternatively, perhaps spongiophytes were lichens consisting of two or more varieties of mycobionts as well as a phycobiont. If the latter, these so-called primitive Devonian organisms may have been far more complex than generally thought.

Even more interesting than the spongiophytes, and certainly more bizarre, are the nematophytes. The best known nematophyte is the genus *Prototaxites* (figure 6.49). The nature of

Prototaxites is one of the great unsolved mysteries of paleontology.

The nematophyte *Prototaxites* was first described by J. William Dawson in 1859 from Siluro-Devonian rocks, again from the Gaspé. Dawson named the fossil *Prototaxites* because he thought that its microstructure resembled the tracheids of the modern yew, *Taxus*, but it is now known that *Prototaxites* has no close relationship to yew or indeed to any vascular plant. Like some vascular plants, however, *Prototaxites* attains considerable size; specimens as much as one meter in diameter and up to four meters long have been reported.

Because of its large size, this genus has generated considerable controversy over the nature of the organism responsible for the fossil remains.[37] Large specimens resemble fossil tree trunks, although *Prototaxites* fossils do not undergo as much compaction with burial as is often the case for fossil wood. Despite its resemblance to a tree trunk, there is no direct evidence that *Prototaxites* had roots. Its large size, however,

FIGURE 6.49

Prototaxites log. View of a large specimen lying parallel to the bedding surface of tilted sediments in a road cut near Cross Point, Quebec. No *Prototaxites* log has ever been found erect.

argues for a land habitat. If *Prototaxites* was similar to living aquatic algae (such as kelp) it is difficult to understand how it could have survived with a meter thick "trunk" completely immersed in water, without the core being rotted out by anaerobic bacteria. Its filamentous microstructure *is* reminiscent of fungal hyphae, but if *Prototaxites* were related to fungi, it is difficult to figure how the organism could have obtained sufficient food to maintain the cells in its enormous girth unless it was photosynthetic.

Prototaxites looks, in thin section and under high magnification (figures 6.50–6.53), like a giant bundle of hollow, con-

FIGURE 6.50 (Left)

Prototaxites in thin section, cross section through log, showing the filaments and larger tubes typical of the genus. Diameter of individual tubes is 40 to 50 microns. Specimens in this and the following three pictures are from the Upper Devonian of Ontario.

FIGURE 6.51 (Right)

Prototaxites in thin section, this time showing a longitudinal section through log. The tubes are straight, compared to the twisted masses of filaments. Diameter of individual tubes is 20 to 30 microns.

 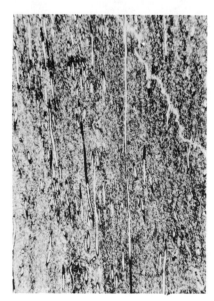

torted spaghetti noodles packed around a lesser number of more-or-less straight drinking straws. In other words, there are two types of elongate structures forming the meter thick "trunk" of *Prototaxites*. The smaller (spaghetti) elements are called filaments (or hyphae) and are about 5 to 10 microns in diameter with thin walls; the larger (straw) elements are called tubes and are about 20 to 50 microns in diameter with thick walls. In some *Prototaxites* specimens, some of the filaments grade into larger tubes (William E. Stein, personal communication).

The tubes branch occasionally and can constitute up to half of the *Prototaxites* tissue. Each tube is enveloped by a tightly packed sheath of filaments. The filaments have cross walls or septa that are well separated—a tenth of a millimeter apart—with a complex elliptical pore perforating each septum (figure 6.53). This pore is reminiscent of the dolipore septum of a living basidiomycete fungus. The filaments in *Prototaxites* may have been coenocytic, that is, they may have shared a continuous mass of cytoplasm, containing several nuclei and formed

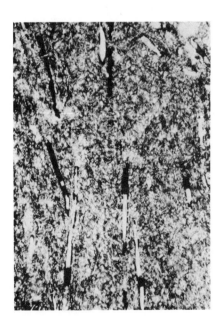

FIGURE 6.52

Another view of *Prototaxites* in a longitudinal thin section, enlarged to show the relationship between tubes and filaments. Diameter of individual tubes is 30 to 50 microns.

from an original cell with one nucleus, much like a basidio-mycete.

The collar-shaped pore in the filaments of *Prototaxites* shown in figure 6.53 would render the filaments ineffective as water transport tissue. One might thus hypothesize that the tubes, not the filaments, were the water transport tissue. They would then be analogous to the tracheids or conductive cells of xylem in a familiar vascular plant. The filaments, then, would be analogous to phloem. Perhaps the odd-looking septal pore in the filaments represents a microvalve, a sort of internal stomate, "bioengineered" by natural selection to regulate counterflow of nutrient-rich water.

No leaflike organs have been described from *Prototaxites*, although many researchers do seem to assume that it was photo-

FIGURE 6.53

Septum of a *Prototaxites* filament with elliptical septal pore. The function of this pore is unknown, but it is clear that a septum with this half-blocked pore could not have been very effective for transporting water. The filament would have been about 5 to 10 microns across. Scale bar = 4 microns.

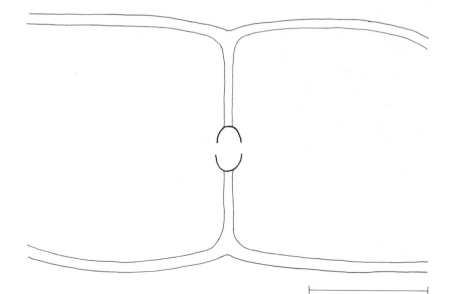

synthetic. A look at other nematophytes seems to support this assumption. A related but much smaller genus, with the typical filament-rich sort of stalk, also has cuticular leaflike appendages. Paul Strother has recently discovered a small, as yet unnamed, branching nematophyte (figure 6.54).

Nematothallus (figure 6.55) seems to show a direct connection between the filamentous tangle and the cuticular portion of the fossilized leaflike organ. Cuticle-covered organs of this sort are almost certainly of terrestrial origin, since cuticle this thick would inhibit photosynthesis under water. Intriguing reports have been made of clusters of dyad cryptospores occurring in association with *Nematothallus*, perhaps indicating that nematophytes had dyadic spores.

Discerning the form and function of *Prototaxites* is a valuable endeavor in itself, but our interest here is Hypersea—specifically the water transport capabilities of the vascular plants that are the body and soul of Hypersea. If *Prototaxites* transpired water at a rate similar to that of a modern tree with similar trunk diameter, how much plant could it keep moist with this piping system? To rephrase the question, a large

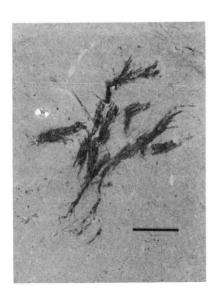

FIGURE 6.54
Fossil of a branching nematophyte from the Rose Hill Formation, Monterrey, Virginia. Early Silurian, late Llandoverian Age. Scale bar = 5 mm.

Prototaxites "trunk" could supply water to a plant of what size?

For comparison, we will use the alpine sorrel *Oxyria digyna* because of its tolerance of a very harsh environment and because of its unregulated stomates. This small angiosperm plant with kidney-shaped leaves is a relative of the common weedy dock and buckwheat. It grows in difficult high-altitude environments where snow does not melt until midsummer. The plant is found worldwide in northern latitudes. The ecology of *Oxyria* has been best studied where it grows on scree (bare, broken rocks forming a sheet of coarse debris) at an elevation of nearly 2,800 meters in the Sierra Nevada of California. Seedling establishment is tenuous in this alpine environment, and it is strictly limited by the availability of liquid water.

Oxyria is an obligate "snow bank" species in the Sierra, because it apparently has no built-in control of transpiration. Early land organisms such as the spongiophytes, which had pores without true stomates, may also have had no built-in control of transpiration. Recall that most modern plants are able to regulate water loss from leaves with the help of stomates and the guard cells surrounding them. When water is plentiful, the guard cells are turgid and the pores open; under conditions of drought, the guard cells loose their turgor, become flaccid,

FIGURE 6.55

Nematothallus compression fossils. A. Small flattened tubes of *Nematothallus*. Scale bar = 100 microns. B. Transmitted light photograph of an acetate peel of axial specimen of nematophyte, which appears to be attached to a tubular thallus of *Nematothallus* sp. Width of tubular thallus is 4.7 cm.

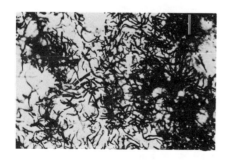
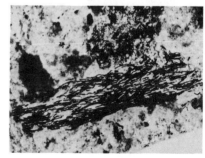

and close the stomates, thus limiting water loss from the interior spaces of the leaves.

The stomates in *Oxyria* always stay open, and its rate of transpiration rises in concert with increases in evaporation from its leaves. On a sunny afternoon *Oxyria* will loose 3 grams of water per 100 square centimeters of leaf area. The plant is somehow able to obtain sufficient water for this unchecked transpiration.[38]

Both *Oxyria* and *Prototaxites* are organisms adapted to marginal or stressful environments. Both probably need or needed to grow near sources of melting or standing water, respectively. The modern *Oxyria* is stressed by being at or above the timberline;[39] the Devonian *Prototaxites* could be considered a pioneer organism testing out the difficult transition to land life. With regard to evaporation stress and lack of stomatal regulation, there may be some interesting similarities between the lifestyles of *Oxyria* and *Prototaxites*. Assume, for the sake of the calculation to follow, that *Prototaxites* had pores without functioning guard cells, and that it had a surface areal rate of water loss through transpiration approximately equivalent to that of *Oxyria*. Some early Devonian spongiophytes like *Orestovia* had stomate-like pores on the thallus, but nothing resembling surface stomates have been found in nematophytes. An alternative interpretation would be to have the filaments of *Prototaxites* be the xylem analogs, in which its odd septal pores (figure 6.53) might play a role in regulating the flow of transpired water. It is difficult to imagine how they might have functioned as stomates, however.

Utilizing the *Oxyria* data for *Prototaxites* will provide only a very rough estimate of *Prototaxites* transpiration but is the best one possible at the moment. Assume that *Prototaxites* had multiple photosynthetic surfaces or a single gigantic photosynthetic surface which connected directly to the trunk. The latter case might at first seem very unlikely until one considers the unusual early Devonian fossil *Sporogonites exuberans*, known from Belgium. This curious fossil consists of a sheetlike surface fifteen centimeters by five centimeters that is studded with four-centimeter-high branches bearing spore sacs. This fossil is clearly only a fragment of a much larger whole.[40]

As a fossil, *Prototaxites* is always found prostrate on its side; it has never been found upright, so we assume that the horizontal position was its growth position. This assumption may not be correct; some paleobotanists think that *Nematothallus* was probably an upright stem, perhaps with an unpreserved mushroom-shaped cap above. Nevertheless, because of the absence of vertical specimens, we will use the horizontal growth position for our model. Using the assumptions in the two preceding paragraphs, *Prototaxites* can be modeled as a living tree on its side. One end was rooted firmly in the water or soil and the other was raised slightly higher in the air. Consider a trunk 1 meter in diameter. Thirty percent of the tissue of the trunk consists of conductive tubes averaging about 35 microns in diameter. We calculate that there were 245 million of these thin tubes running the length of the *Prototaxites* trunk.[41]

Using *Oxyria* transpiration data, we assume that loss of water from the thallus of *Prototaxites* is 0.03 g/cm^{2}hr. Next we'll calculate an estimate for a rate of transpiration for a tree with a trunk one meter in diameter, and use that as an estimate of the amount of water transpired by the hypothetical stomateless photosynthetic organs of *Prototaxites*. Combining this with the *Oxyria* data, we will be able to calculate the area that the massive *Prototaxites* trunk could supply with water. We will also assume that *Prototaxites* trunks had to be so thick because the "leaves" lacked guard cells and thus lost water at a relatively high rate.

The transpiration rate of trees in tropical rain forest in the Amazon Territory of Venezuela has been measured using tritiated water. The radioactive form of hydrogen in tritiated water provides a tracer. It was possible to show that transpiration is a linear function of the sapwood area of the trees measured. Sapwood is the part of the wood that actively transports fluids and can store nutrients; heartwood has lost the ability to store nutrients and transport substances, but has a high degree of mechanical self-support.[42]

A graph of transpiration rate versus sapwood area plots as a straight line. The regression coefficient (how well the data fits a straight line) of the Venezuelan tree data is 0.96, a very high value (perfect correlation is 1.00) indicating a close fit of the data to the regression line. Applying the regression line formula to the one meter trunk diameter of *Prototaxites*, we calculate that the trunk could supply with water a prostrate photosynthetic organ with a surface area of nearly 300 square meters. This is large but perhaps not unreasonably so considering the relative lack of terrestrial herbivores and competitors during the Devonian and considering that flat prostrate organisms like *Sporogonites exuberans*, mentioned earlier, also are known from the Devonian. Why the hypothetical *Prototaxites* thallus did not fossilize is a mystery; perhaps it was easily disaggregated after death. Small loose shreds thought to be nematophyte thallus are known from the Rhynie Chert (the nematoclasts shown in figure 6.9).

Of course, our assumption that *Prototaxites* had a giant, lost photosynthetic organ could well be wrong. Recall the spongiophytes that may have possessed more than one type of mycobiont. By comparison with spongiophytes, perhaps *Prototaxites* was a giant lichen, perhaps even a mycobiont-mycobiont lichen lacking a phycobiont. A fungus does not necessarily require a phycobiont to grow large. The modern *Armillaria bulbosa*, the honey mushroom, exhibits an enormous subterranean sprawl. *Armillaria* does not need an internal photosynthesizer to grow to great size; perhaps the same was true of *Prototaxites*. The filaments (thin hyphae?) in *Prototaxites* can be shown with microscopy to cluster tightly around the tubes (thick hyphae?). Perhaps this indicates some type of nutrient exchange between the two different sizes of tubular elements. Feeding may have involved absorption of reduced compounds or other nutrients, Garden of Ediacara style, from loose sediments or soils. *Prototaxites* may have been a member of a "Garden of Ediacara of the Land."

Why did *Prototaxites* so soon fall out of the evolutionary story? Perhaps is was less biohydraulically efficient than a vascular plant. Perhaps it *was* unable to photosynthesize, and ran out of adequate food supplies for some reason. Perhaps it was eaten by some kind of newly evolved animal. Perhaps the genus was just unfortunate during a time of mass extinction.

In any case, *Prototaxites* probably represents an ultimately unsuccessful attempt to colonize the land by large, algalike or funguslike organisms; organisms which in a sense can be thought of as a failed evolutionary experiment. We consider it quite likely that spongiophytes and *Prototaxites* each represent lichenlike consortia of two or more types of unrelated organisms, bound together in a Hypersea linkage. If so, these organisms can be considered (at least ecologically) to have left a successful living legacy in the form of modern lichens, which still dominate the land surface of a number of habitats. The "weird" aspects of spongiophytes and *Prototaxites* may have to do with their possession of unfamiliar types of hypermarine symbioses, as for example a lichenlike association of not one but two mycobiont fungal partners.

The Origin of Plant-Fungal Symbiosis

The oldest reasonably diverse fossil fungi are from the 400-million-year-old (early Devonian) Rhynie Chert, a rock formation near Aberdeen, Scotland. The Rhynie chert (figure 6.56) was discovered early in this century by a geologist making a geologic map of the area. The locality cannot be viewed in outcrop today because, as noted earlier, the bedrock exposures have been buried and turned to cow pasture.

The deposit is best known for its superb preservation in silica of the early vascular plants *Rhynia* and *Asteroxylon* (figure 6.20), nonvascular plants such as *Aglaophyton* (figure 6.1), nematophytes (figure 6.9), tiny arthropods such as the crustacean *Lepidocaris* (figure 6.57), and predatory arthropods, such as the fanged trigonotarbids (figure 6.58). As of this writing, trigonotarbids are the oldest land arthropods known. Most of the fossil plants have been interpreted as sporophytes, but evidence suggests that there may also have been gametophytes comparable in structure and size to the sporophytes.[43]

From their habitat and anatomy it appears that these early Devonian plants primarily relied on turgor rather than on structural stiffening of stems for support. Unlike *Prototaxites*, the Rhynie plants were small enough for turgor support to be effective. All the plant genera at Rhynie seem to be endemic to the locality; they have been found nowhere else. But this endemism may be only apparent. Perhaps the exquisite three-dimensional preservation of these plants masks their affinities to compression fossils (flattened in shale) at other localities. For instance, paleobotanists suspect that the Rhynie plant *Aglaophyton* may be identical to the carbonized compression fossil *Salopella*. Nevertheless, the Rhynie Chert flora may have been a highly specialized community adapted to existence in the vicinity of a hot spring environment in which the fossils were preserved. The Rhynie plants are morphologically odd (especially the water-conducting system of *Aglaophyton*) and do not compare closely to modern plants.

The idea that the simple structure of the Rhynie plants owed to special environmental conditions was proposed by Edward

FIGURE 6.56

A chunk of the Devonian-aged Rhynie Chert. This Scottish rock formation contains the earliest diverse fungal biota and the oldest known direct evidence for the existence of Hypersea. Width of specimen is 7.5 cm.

W. Berry in 1945. Berry derided as antiquated the notion that the Rhynie plants were "primitive," and he argued that "antecedent and contemporary plants were on a much more complex structural level. The bog plants are discordant with the stage of evolution they have claimed to represent. Also discordant in size." This view has gained recent support, and arguments have been presented to the effect that it is impossible to relate the rhyniophytes to any group of putative algal ancestors or to any group of putative vascular descendants because they represent a "blind alley" in early land plant evolution.[44]

Recent reports of fossil land plants from the latter part of the early Devonian in the Gaspé region of Quebec depict plants whose more complex anatomy indicates a level of stem-leaf differentiation significantly more advanced than previously known for trimerophytes or other vascular plants of this age.[45] The Gaspé fossils thus suggest that more "normal" plants were contemporaries of the Rhynie plants. This would accord with Berry's view that the Rhynie plants were unusual because they were adapted to special conditions.

In spite of the unique fossil plants, the most interesting features of the Rhynie Chert biota are the abundant fossils of fungi. In chapter 1 we introduced the Rhynie Chert as being

FIGURE 6.57

Lepidocaris, a polyramous crustacean from the Rhynie Chert. A. photomicrograph showing five leg branches. Scale bar = 25 microns. B. enlargement of A, showing details of limbs. Note numerous branches on each leg. Scale bar = 5 microns.

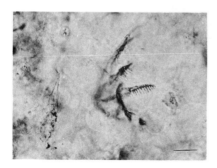
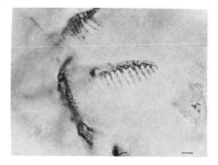

important for its nested fossils—specifically, a fossil fungus occurring within an alga that, in turn, was within a fossil Rhynie plant (figures 1.2 and 1.3). After decades of neglect, these and other Rhynie fungal fossils are finally attracting the attention they deserve.

All these fungi were preserved in silicified peat along with the plants. The ancient spring or wetland was inundated by hot, silica-charged waters, perhaps a result of periodic pulses of hydrothermal fluid. The local environment may have had much in common with the hydrothermally active areas of Yellowstone National Park.

As the hydrothermal waters spilled onto the surface and cooled, the silica formed tiny crystals, nearly invisible even under high magnification, thus producing what is called cryptocrystalline chert. In these fossiliferous cryptocrystalline cherts, fragile details of many microscopic organisms are preserved, providing a biotic "snapshot" of life in the Rhynie hot spring. The fungi and their spores or resting cysts are mostly found in the stems and rootlets of the plants (figure 6.59). One fungus-covered arthropod specimen found in the chert has a large fungal chlamydospore in its body cavity.[46] A chlamydospore, also called a resting spore or a vesicle, is an asexual spherical structure of fungi that forms by the swelling and partitioning of a living hyphal segment. In modern fungi, the structure formed is used for survival during seasonal desiccation.

Modern mycorrhizal fungi form chlamydospores very similar to those seen in the Rhynie Chert. Mycorrhizal fungi are,

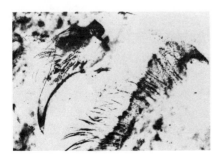

FIGURE 6.58

Fang (chelicera) of a Rhynie trigonotarbid, one of the earliest terrestrial arthropods known. Trigonotarbids are arachnids, relatives of spiders. The trigonotarbid held its claw under its body, and presumably used it for subduing small prey. Length of fang is approximately 400 microns.

according to Hypersea theory, the most important organisms on land. These fungi, such as the mycorrhizal genus *Glomus*, form tight symbiotic linkages with the rootlets of vascular plants. The plant rootlets swell and expand with internal fungi. As will be discussed below, this quintessential hypermarine partnership underpins most of Earth's biomass. For well over a hundred years, it has been recognized that many vascular plants live in close association with mycorrhizal fungi, but their importance is generally overlooked.

In addition to chlamydospores, modern mycorrhizae also form spores within spores (of the same or a different species) by a process called internal proliferation. We have found evidence in the Rhynie Chert of internal proliferation (figures 6.60 and 6.61) quite similar to the internal proliferation of spores in the modern mycorrhizal genus *Glomus*.[47] This similarity, together with the chlamydospores in figure 6.59, suggests that mycorrhizal fungi belonging to *Glomus'* family (Glomaceae), or to a fungal family ancestral to the Glomaceae, are present in the Rhynie Chert.

Rhynie Chert fungi do lack one important feature of modern mycorrhizal fungi. Arbuscles are three-dimensional, branched

FIGURE 6.59

Fungal resting cysts are preserved within the stem of a Rhynie Chert plant. A. a cluster of several cysts; B. enlargement of a single cyst. Cysts are approximately 25 microns in width. Chert, a type of rock composed of submicroscopic crystals of the mineral quartz, is able to preserve microscopic features in amazing detail.

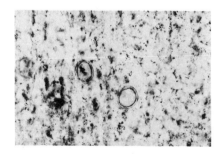
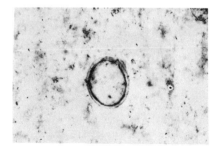

bush-shaped structures formed by fungal hyphae *within* the root cells of plants. They are thought to be involved in nutrient transfer in the symbiotic relationship between fungi and plants, as in the living lycophyte *Lycopodium cernuum* that we discussed earlier. Fungal hyphae insert themselves into a plant root cell to form the specialized branched arbuscles. The branching probably increases the surface area available for nutrient exchange between the fungus and the plant, and it is reminiscent of the finely branched hyphae of the lichen mycobiont that intrude into the algal or cyanobacterial phycobiont.

The oldest fossil arbuscles are Triassic in age, from ancient silicified roots found in the Antarctic.[48] Arbuscles are completely absent in the Rhynie Chert. This fact suggests to us that the Rhynie Chert was deposited during the evolutionary beginnings of the mycorrhizal fungal habit, and that arbuscles

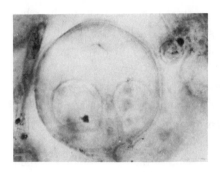

FIGURE 6.60
Glomus-like fungus from the Rhynie Chert. A spore body containing smaller spore bodies; two internal spore bodies are visible at this plane of focus, a third is present above the plane of focus. Scale bar = 5 microns.

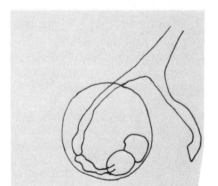

FIGURE 6.61
Drawing from a photograph of internal proliferation of *Glomus*-like spore capsules of a Devonian fungus. One internal spore body presses against a second internal spore body, on its right. Scale bar = 10 microns.

were a later evolutionary development, possibly one that enhanced the efficiency of nutrient exchange between the plant and the fungus. The presence of *Sclerocystis* or a related form of mycorrhizal fungus (figure 1.4) in the Rhynie Chert also suggests that more than one genus of fungus was testing the mycorrhizal relationship. This is yet another example of convergent evolution. *Sclerocystis* is an important mycorrhizal genus today.

Experiments conducted on seedlings of the conifer *Pinus strobus* demonstrated that in nutrient-deficient soils, seedlings with mycorrhizal symbionts show increases in weight and disproportionate increases in nutrient absorption when compared to non-mycorrhizal control seedlings. The mycorrhizal seedlings were 26% heavier, had 45% more dry weight nitrogen, 150% more dry weight phosphate, and 72% more dry weight potassium than did their non-mycorrhizal counterparts.[49]

Mycorrhizal fungi secure food from the roots of their host while contributing substantially to the host plant's absorption of water and mineral ions, particularly in plant species with limited numbers of root hairs. In sudangrass (*Sorghum vulgare*), a plant with mycorrhizal fungi, plants that were experimentally provided with extra phosphorus showed an overall decrease in mycorrhiza formation, suggesting a link between mycorrhiza quantity and plant nutrient requirements.[50]

Such facts may explain why more than 80% of living plants have obligate mycorrhizal associations: they cannot live without their fungal partners. In fact, all of the basic types of plants are mycorrhizal; 90% of living plants have mycorrhizae of one sort or another (though not all are obligate). The famous French delicacy truffles, for instance, are actually the reproductive structures of the mycorrhizae partnered with hardwood trees. Modern plants are thus, for all practical purposes, composites of a plant "phycobiont" and mycorrhizal fungi.

In at least one case the mycorrhizae are so adept at securing nutrients that the associated plant need rely on no other source, not even photosynthesis, for food. The orchid *Corallorhiza innata* is parasitic on its mycorrhizal fungi. This colorless orchid does not photosynthesize and has a reduced rooting system with such a heavy mycorrhizal association that it exhibits thick mantles of mycorrhizae around the roots. The fungi are able to absorb sufficient food to support both themselves and the freeloading orchid. Many other orchids are so dependent on their mycorrhiza that they cannot germinate until the seed coat has been partially decomposed by a potential fungal partner. This requirement explains why some orchids are extremely difficult to germinate.[51]

The Fungal Fusion Hypothesis of Pirozynski and Malloch

There is evidence for close association between fungi and vascular plants in the Devonian in both the submerged or buried parts and the aerial parts of the fossil plants. It is not hard to imagine mutualistic interactions between plant and fungi being highly favored in the early evolutionary stages of such situations. Any evolutionary change that helps the plant get more nutrients from the soil can help the symbiotic fungi get more nutrients from the sap. This sort of thinking, taken to its logical extreme, has led to the Fungal Fusion hypothesis.

Promoters of the Fungal Fusion hypothesis (proposed in 1974 by K. A. Pirozynski and D. W. Malloch) argue that an even more intimate relationship led to the symbiogenetic synthesis of plants in the first place.[52] In other words, fungi and a photosynthetic algal symbiont joined forces to form the first plants. Supporters of the Fungal Fusion hypothesis see land plants as complex, overgrown lichens with no clear distinction between phycobiont and mycobiont.

The Fungal Fusion hypothesis claims that vascular plants are actually the result of a symbiotic fusion of a green alga and a fungus. This idea is carried even further in a 1991 article entitled "Fungi and the Origin of Land Plants" by Peter R.

Atsatt.[53] Atsatt expands on the Fungal Fusion hypothesis by claiming that vascular plants are actually a fusion, in both a somatic and a genetic sense, of fungi and green algae. Land plants, he notes, display a combination of both the photosynthetic mode of nutrition utilized by algae and the degradative and absorptive mode of nutrient acquisition typical of fungi.

The ability to secrete wall-degrading enzymes called chitinases in a controlled and localized manner is typical of both fungi and higher plants when they are developing complex tissues.[54] The acquisition of funguslike degradative and absorptive nutritional mechanisms in vascular plants could, of course, be an example of convergent evolution, but Atsatt stresses the possibility that these land plant traits may be clues to the incorporation of fungal genes into the plant genetic code as a result of a single symbiotic event or by transfer of genes from fungus to the alga that gave rise to plants. This idea of horizontal gene transfer or lateral gene transfer, once anathema in polite scientific circles, has been decisively demonstrated in a number of cases involving unrelated organisms.

Evidence of Fungal Fusion   Evidence supporting the Fungal Fusion hypothesis includes the existence in plants of transfer cells, which are specialized for transfer of nutrients between hypermarine-linked organisms. Transfer cells are also characteristic of the interface between the gametophyte and the embryonic sporophyte in the green alga *Coleochaete* (figure 6.43), and they often develop at the interface between root and fungus in modern mycorrhizal associations. As further evidence, the endosperm in the seed of the tree *Nothapodytes* produces branched haustoria that invade nearby cells of the parent plant, causing parent nuclei to degenerate. Aggressive haustorial cells that penetrate host cells and absorb nutrients are characteristic of parasitic fungi.

Tubular haustorial growth is also exhibited by female gametophytes of flowering plants. Even plant embryos show funguslike characteristics. Atsatt contends that the borrowing of a fungal gene for extracellular digestion and absorption led to the evolution of embryos in early plants. Atsatt calls these "selfish embryos." They act as parasitic sporophytes capable of eliciting the release of maternal nutrients in a manner not unlike that of a parasitic fungus.

The haustoria of *Gnetum* embryos aggressively parasitize their own endosperm (the "sister sporophyte" which provisions a seed with food) and sometimes continue their foraging into the maternal sporophyte. Sometimes sporophyte plants actually parasitize sporophytes of other species. Achlorophyllous parasitic plants such as dodder (*Cuscuta*) are one example, and there are even plants that live hyphalike within photosynthetic tissues of other plants.

Pollen tubes, the largest organ formed by the miniaturized male gametophyte in angiosperms, have a very funguslike development. This tiny male gametophyte is able to invade sporophyte tissue like a parasite in order to fertilize the female gametophyte; the acquisition of this ability by Paleozoic seed fern pollen was a revolutionary event that permitted the origin of seeds and flowering plants. Pollen tubes have morphological similarities to hyphae that suggest to proponents of the Fungal Fusion hypothesis a relationship between pollen and fungi that is not mere evolutionary convergence.

Atsatt's hypothesis has gained recent support. In a 1993 paper Richard Jorgensen noted that the hyphalike character of actively growing pollen (attained by elongate tip-growth of funguslike cells) has been shown (in *Petunia* pollen) to be promoted by a class of chemicals called flavonoids.[55] Flavonoids are substances that protect plants from damage by ultraviolet light. In the same way that endosymbiosis may have once sheltered tiny marine organisms within the tissues of larger organisms during the Oxygen Revolution, Jorgensen suggests that a flavonoid-synthesizing alga became colonized by a tip-growing, funguslike organism that entered the algal tissues, at least in part because such tissues could act as a refuge from the destructive effects of ultraviolet light on land. Symbiosis and horizontal gene transfer eventually followed. Jorgensen thus hypothesizes that land plants were born as a synthesis of a flavonoid-bearing green alga and an invasive fungus. If Atsatt and Jorgensen are right, land plants have a hybrid set of genes.

The evidence that fungi and plants share gene complexes includes striking similarities in the presence of wall-degrading enzymes, presence of haustoria, and parasitic habits. Both Atsatt and Jorgensen suggest that to test the Fungal Fusion hypothesis and to rule out the possibility of convergent evolution, one should begin with molecular biological study of wall-degrading enzymes such as β-1,3 glucanases and chitinases and the vacuolar and plasma membrane ATPases. Even if the similarities do turn out to be merely due to convergent evolution, the Fungal Fusion hypothesis is fascinating because it compels examination of the evolution of early plants from a perspective that emphasizes their funguslike properties. More importantly, it highlights the tendency for one type of terrestrial eukaryote to live inside of another on land.

This tendency surely has marine antecedents at the cellular level, as discussed earlier when we considered symbiogenesis, the Oxygen Revolution, and the marine origin of the eukaryotic cell. The Rhynie Chert provides a fascinating example of organisms within other eukaryotic organisms, an example that gives a penetrating insight into the origins of Hypersea. As one might expect, parts of Hypersea originated in freshwater aquatic environments.

Charophytes, also known as stoneworts (*Chara*) or brittleworts (*Nitella*) are a type of green alga closely related to *Coleochaete* and to plants. They characteristically live in freshwater lakes and streams, and they have a superb fossil record extending back to the Devonian. The reason for this excellent fossil record is that charophytes precipitate calcium carbonate around their bodies, thus enhancing the likelihood of their preservation as fossils. The oogonia (or embryo sacs) of charophytes are particularly resistant to decay and constitute the majority of charophyte fossils. The oogonia are clearly desiccation-resistant structures, testimony to the intermittent wet/dry environment of charophytes. Charophyte oogonia are rounded to oval, with spiraling grooves running from pole to pole, giving them the appearance of tiny twisted footballs about four tenths of a millimeter long. Devonian charophyte species have euphonious names such as *Trochiliscus surmilovae*, *Umbella bella*, *Umbella bashkirica*, and *Quasiumbella nana*.[56]

Charophytes are represented in the Rhynie Chert by the species *Palaeonitella cranii*. The genus name emphasizes the similarity of this fossil to the modern brittlewort *Nitella*, a common pond weed with rings of branches that look like grass skirts clustered at regular intervals (nodes) along its main stem. There are no branches in the smooth, internodal sections of the *Nitella* or *Palaeonitella* stems.

Palaeonitella provides a fascinating example of ancient parasitism because specimens of the Rhynie brittlewort are preserved with evidence of infection by a funguslike protoctist belonging to the phylum Plasmodiophoromycota, whose members are all obligate parasites. Plasmodiophoromycetes are a common but inconspicuous group of intracellular parasites with two distinct parasitic stages in their life cycle. One reason that they are poorly known to biologists, even though they associate with a great diversity of different types of organisms, is that they must always be grown on or in their hosts. Plasmodiophoromycetes are usually plant parasites, and they are responsible for causing the club root disease of cabbage. On the other hand, plasmodiophoromycetes are often asymptomatic in their normal hosts. Many plasmodiophoromycetes must be cultivated on Chinese Cabbage (*Brassica pekinesis*) in order to ensure a visible infection in this less-resistant plant. The plasmodiophoromycete genera *Ligniera* and *Polymyxa*, found in root hairs and roots of their hosts, must be collected from roots searched microscopically because they do not cause any swelling of host cells.

Fossils of swollen or hypertrophied internodal cells in

Palaeonitella indicate a diseased state very similar to the host response to plasmodiophoromycete invaders in modern charophytes. An infected *Paleonitella* specimen has a bulge in the smooth internodal section like a garden hose ready to rupture; it is identical to a modern plasmodiophoromycete-infected *Nitella*. In addition, one *Paleonitella* specimen shows the plasmodiophoromycete in the process of rupturing the charophyte cell wall with its pressurized cyst. When the cell wall ruptured, the plasmodiophoromycete transferred its spores to the interior of the alga. This recently discovered association in the Rhynie Chert also establishes the first appearance of plasmodiophoromycetes in the fossil record.[57]

In addition to being parasites of chlorophytes, plasmodiophoromycetes today are hyperparasites of numerous types of fungi. The plasmodiophoromycete *Sorodiscus* invades both charophytes and oomycete fungi such as *Pythium*. The plasmodiophoromycete *Octomyxa* invades hyphae of the oomycete fungi *Achlya*, *Brevilegnia*, and *Geolegnia*. As parasites and hyperparasites, plasmodiophoromycetes play an important role both in aquatic environments and in Hypersea.

What we see in the Rhynie plasmodiophoromycetes is an example of a biological interaction that was a possible aquatic precursor to hypermarine interactions. A desiccation-adapted green alga was invaded by plasmodiophoromycetes, a phylum which may have originated as part of the first terrestrial aquatic ecosystems. These plasmodiophoromycetes became well suited to endoparasitism, and subsequently spent most of their lives within the tissues of other eukaryotes. As chlorophyte relatives moved into truly terrestrial environments, the plasmodiophoromycetes followed.

Although mostly considered to be disease organisms today, plasmodiophoromycetes in the roots of many plants are apparently asymptomatic and may be benign members in a root-dwelling assemblage of unrelated organisms—or, if you will, a Hypersea community.

The Case of Vigorous Tall Fescue   Not only are some Hypersea organisms asymptomatic, but others are quite beneficial for the host plant. The well-known example of mycorrhizae was discussed earlier in this chapter. Another plant-fungus association, this time penetrating the upper photosynthetic part of the plant, has proven so successful that it is, ironically, causing severe agricultural and animal husbandry problems in the United States at the same time that it helps the host plant thrive.

Bucolic landscapes of many parts of eastern and central North America have a rich green hue due in large part to the tall fescue grass (*Festuca arundinacea*) so common in these regions. Tall fescue comes in a number of varieties, the most vigorous of which was widely planted throughout the eastern half of the United States from about 1943 to 1960. The variety known as vigorous tall fescue (VTF) was first discovered in 1931 by E. N. Fergus of the University of Tennessee. Fergus found the grass on a hillside on the Menifee County, Kentucky, farm of William O. Suiter. Fergus noted that the Menifee ecotype of fescue was particularly vigorous, and after seed was evaluated and passed as suitable for agricultural purposes, Fergus released it to the farming community as the fescue cultivar Kentucky 31.

VTF is today, by an overwhelming margin, the most common type of tall fescue; the grass now covers about 35 million acres of the United States. VTF is amazingly hardy and drought resistant. VTF grows in areas where other strains of tall fescue cannot, and it even seems to be able to grow in areas where no other types of agricultural grasses can survive. After its release by the University of Tennessee, it appeared to be a "wonder grass," a triumph of American ingenuity and agricultural research.

But troubling reports soon began to surface regarding the dangers that this tall fescue posed for livestock. Three very unpleasant livestock disorders, reminiscent of ergotism, came to be associated with animals using VTF for forage. One symptom is called fescue foot. In fescue foot, the animal's extremities, including hooves, feet, and the tips of tails and ears, become gangrenous and may actually fall off. Fescue foot is worst in cold weather. A second symptom is bovine fat necrosis. This disease is manifest as lumps of hard fat in the abdominal cavities of cows, bulls, and steers. Third is fescue toxicity, characterized by symptoms such as reduced reproductive capacity, spontaneous abortions or prolonged pregnancy, drop in milk production, slow weight gains, and poor appetite.[58]

Research in the 1960s demonstrated that these three kinds of animal symptoms were brought about by ingestion of an unknown toxin by the animals that had grazed on VTF. The toxin posed no threat for human consumption of affected meat, for it was completely broken down in the animal's system. This fact did not mitigate, however, the economic troubles for livestock farmers faced with the debilitating symptoms caused by this variety of vigorous tall fescue.

The toxicity was traced to an alkaloid compound in the grass. The source of the toxic alkaloid was puzzling at first but it was finally traced to an endophytic fungus living *within* the photosynthetic part of VTF (figure 6.62). This fungus, first thought to be the same fungus that causes choke disease (*Epichloë typhina*) in grasses, was later shown to belong to an undescribed genus. This new genus (and species) was given the name *Acremonium coenophialum*. The genus name reflects the acrimonious effect of the endophytic fungus on mammalian herbivores and the farmers who depend on them.

As shown in figure 6.58, filaments of *Acremonium coenophialum* grow into the intercellular spaces between VTF cells. The fungus is not injurious to the grass—quite the contrary. The vigorous character of this fescue variety is due directly to the presence of the endophyte. The fungus confers drought resistance to the grass, and infected grasses produce more seeds than do uninfected grasses. The infected seeds sprout into larger and healthier seedlings than do their uninfected counterparts. Furthermore, grazing insects and root nematodes seem to shun fescue bearing the endophytic fungus. (Evidently, cattle should follow their example!)

The fungus is carried from generation to generation of fescue by a tiny mycelium. As in a lichen propagule, the threadlike hypha wraps around the grass embryo while it is still completely within the seed coat. Later the seed germinates with both embryo and fungus intact. As the grass grows, the

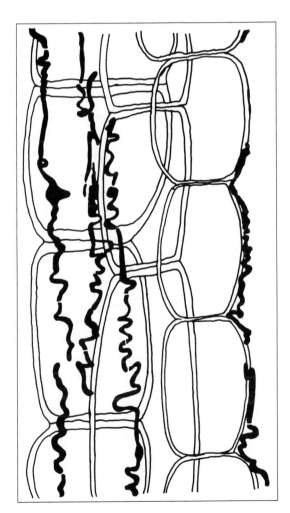

FIGURE 6.62
Fungal hyphae (in black) of *Acremonium coenophialum*, shown colonizing the tissue of vigorous tall fescue. The grass benefits from the symbiotic association, but the livestock that eat it do not. Height of each plant cell is approximately 30 microns.

fungus spreads throughout the green parts of the plant and eventually reaches the grass flower. Each fescue ovule is then infected with a fungal mycelium for incorporation, as an associate of the tiny grass sporophyte, in the next grass generation.

Healthy animals can be raised on pastures of endophyte-free fescue. Farmers have found such fields hard to maintain, however, because the uninfected grasses wither when challenged by lack of water, overgrazing, or other stresses. Fungus-free grass is much less likely to display the leaf-rolling response that helps infected plants preserve water during times of drought.[59]

It therefore seems to us almost as if fungal infestation is the *normal* state for tall fescue and other grasses, and that grasses lacking the endophyte have somehow lost a vital component. It is difficult, however, to see how such a loss could occur if much of the pre-1943 American fescue was fungus-free. But even if the Fungal Fusion hypothesis for the origin of land plants fails to gain widespread support, it is abundantly clear that diverse varieties of fungi can exist within the tissues of vascular plants—from *Lycopodium* to fescue—be those tissues above or below ground.

The hypermarine aspect of plant and fungal evolution can be summed up in a final example. Two closely related liverworts, *Aneura pinguis* and *Cryptothallus mirabilis,* are colonized by fungi. *Aneura* photosynthesizes, but *Cryptothallus,* like some orchids and like the whisk fern (*Psilotum*) gametophyte, is parasitic on the fungi. *Aneura* is able to colonize a wide variety of habitats, whereas *Cryptothallus* seems to be restricted to very specific habitats. This contrast in distribution may have less to do with the fact that one liverwort is photosynthetic and the other is not than it has to do with what is called the fungal specificity of the respective liverworts. *Aneura* has low specificity, which means that it can play host to more fungal symbionts than can *Cryptothallus,* and *Aneura*'s geographic range has expanded accordingly.[60]

This example leads to our hypothesis that the ability to engage in flexible, off-again-on-again symbiosis is a key component for the survival and spread of plant species. In other words, those plants able to make symbiotic associations with a variety of species in their environment will leave more off-spring, and, all other things being equal, will become more widely distributed than the less-flexible species. An early task in any Hypersea research program would be to evaluate this suggestion by determining which plants can enter into symbioses with which fungi and by tabulating compatibilities among the plant and fungus species. We predict that when comparing plants that are otherwise similar, those with the greatest numbers of fungal associates will be the most widely distributed species.

Summary of Hypersea Aspects of Plant and Fungi Evolution

Baragwanathia, the lycophyte of possible Silurian age, is a key fossil for Hypersea studies. Although there is no direct fossil evidence of fungal association with *Baragwanathia*, we suspect (based on its size and similarities to *Lycopodium*) that it was one of the first plants to be associated with fungi. In fact, assuming the Silurian age is correct, the unprecedented early success of this plant may be directly attributable to the influence of fungal symbionts. In our estimation, *Baragwanathia* bears the unmistakable signature of Hypersea. And because the first occurrences of *Baragwanathia* are in Australia, we also suspect that Hypersea originated on or near this continent.

In examining the subsequent evolutionary history of land plants and fungi, we have seen how hypermarine interactions played, and continue to play, a fundamental role in the development and geographic spread of these organisms through geologic time. From the Rhynie fungi to bizarre fossil lichens and fossil mycorrhizal arbuscles, and from plant generations "parasitic" on one another to endophytes in green pastures, we begin to see the overwhelming importance of Hypersea for land plants and their consorts. The fact that these hypermarine interactions are invisible to the casual observer on account of their size or location renders them no less significant. We feel that, as regards the fungal and land plant biota, there is more than meets the eye. In the next chapter we will demonstrate that this consideration also applies to land animals.

SEVEN

Animals and Hypersea

Animal lineages that moved from an aquatic to a terrestrial environment faced challenges similar to those faced by plants. The basic challenge is that the organism is going from a situation of being immersed in water to a situation of being immersed in air.

Terrestrial life presents five special challenges for animals. First, an animal weighs more in air than in water, since air is so much less dense. This causes problems of physical support and locomotion. There are a variety of solutions, most involving either external skeletons, internal skeletons, or both. A bird has both, because it uses wing bones (internal skeleton) and feathers (a kind of soft exoskeleton) to fly.

Second is the water-loss problem. In terrestrial environments other than rainforests, air is hardly ever saturated with respect to water vapor. The threat of desiccation is a formidable problem for terrestrial animals. Water is a small molecule; any respiratory membrane thin enough to exchange carbon dioxide and oxygen will lose water as well. In chapter 5 we explained that some land snails are able to survive desiccation by being osmoconformers capable of tolerating severe changes in the solute content of their body fluids. Insects use a quite different approach: by a still poorly understood mechanism, insect cuticle can actually absorb water from moisture-undersaturated air.

Third is the major problem of food supply in air. Filter-feeding is an extremely common trophic strategy for marine animals, but few animals can filter food from the air. Only web-spinning spiders have much success as terrestrial filter feeders, and even then they have to wait a long time between meals. Thus, predation and herbivory, either or both undertaken while moving from place to place, are the preferred modes of animal nutrition on land.

The fourth problem for terrestrial animals is one of reproduction dependent on water retention. Animal reproductive cells are particularly sensitive to desiccation. Internal fertilization is a common strategy, which is further elaborated by internal gestation and live birth, as in mammals.

The fifth problem is respiration, which is the topic of the following section.

Terrestrial Challenges for Animal Respiration

Major adjustments in animal respiration must have been made before any but the most damp land environments could have been populated. This is so because oxygen and carbon dioxide behave very differently in air than they do in water, particularly as regards their diffusion across respiratory membranes. Animals respire (that is, derive energy from) carbohydrates such as the sugar molecule $C_6H_{12}O_6$ by combining them with oxygen to form carbon dioxide and water. Carbon dioxide is a toxic waste product because it is very soluble in water and forms carbonic acid (H_2CO_3), which can damage cells if its concentration builds up to high levels.

Animals can tolerate small concentrations of CO_2 because their blood is carbonate-buffered, just like seawater. How does this buffering work? It can be easily explained by Le Chatlier's principle of physical chemistry. We provide here an explanation for buffering based on the chemistry of marine water, but for the most part the same formulas are applicable to the body fluids of land animals.

Consider the following chemical reaction formula:

$$A + B \rightleftarrows C + D.$$

The relationship between reactants tends to remain at chemical equilibrium; in other words, if you remove or add quantities of either A, B, C, or D to the system, there will be adjustments in the quantities of the other three reactants. Le Chatlier's principle states that if you add reactant that is on one side of the equation, there will be a shift in composition of the system toward the other side of the equation. Say you have a beaker filled with chemical reactants A, B, C and D, where the reactants have reached equilibrium. Now add extra C. Some of it will combine with the D to create extra A and B.

The carbon dioxide cycle in seawater is the most complicated cycle of all the dissolved marine gases. Pure carbon dioxide is present in seawater only in very small amounts, but it enters into a variety of compounds which permit a vast increase in total aqueous CO_2 concentration. Carbon dioxide occurs in seawater in four forms: as a dissolved gas, as carbonic acid (H_2CO_3), and as dissolved ions of both carbonate (CO_3^{2-}) and bicarbonate (HCO_3^-). (The superscripted "2 − " describes the double negative charge on an ion of carbonate.) The exchanges between carbon dioxide gas, carbonic acid, carbonate ion, and bicarbonate ion can be described as follows:

$$CO_2 + H_2O \rightleftarrows H_2CO_3 \rightleftarrows H^+ + HCO_3^- \rightleftarrows 2H^+ + CO_3^{2-}$$

Add carbon dioxide to the system, and the equilibrium moves to the right: more H_2CO_3, HCO_3^- and CO_3^{2-} form to maintain system balance. Carbon dioxide becomes associated with cations (positively charged ions of hydrogen) to form highly soluble bicarbonates. The formation of carbonate shells and skeletons (actually composed of calcium and/or magnesium carbonate) removes H^+ from bicarbonate and replaces it with another cation to form, for example, $CaCO_3$—which is nearly insoluble except in very cold, deep marine water. In turn, water ionizes to yield more H^+ and bicarbonate ion. If H^+ is added (as through, say, acidic precipitation), bicarbonate reacts with it to store it as carbon dioxide and water. Only trace amounts of CO_3^{2-} ions are present in seawater.

This relation buffers the pH of seawater within the range 8.1 to 8.3 in shallow open ocean water. Once again, if one adds H^+, it drives the system to the left as follows:

$$H_2CO_3 \rightleftarrows H^+ + HCO_3^- \text{ (carbonic acid} \leftarrow \text{hydrogen cation +}$$
$$\text{bicarbonate)}$$

or, at higher pH values,

$$H^+ + HCO_3^{-2} \rightleftarrows 2H^+ + CO_3^{2-}$$

The carbon dioxide concentration is not altered, and the pH remains the same, because the system can absorb the excess acidity (addition of H^+). Lastly, if one adds OH^-, the reaction is driven to the right as follows:

$$OH^- + H_2CO_3 \rightleftarrows HCO_3^- + H_2O$$

and the amount of H^+ is unaffected. This, then, is the buffering of seawater: a tendency to keep the pH of the system constant in spite of additions of acid or base.

Ultimately for animals, however, most of the carbon dioxide generated by respiration must be passed out of the body. Carbon dioxide removal is easy for animals living in seawater, for the gas diffuses away from the animal's tissues very readily. Also, surface seawater is usually well oxygenated; gills can easily pick oxygen up from the water. In air, however, the system of ridding oneself of excess carbon dioxide must be different; gills are useless in air because gill filaments tend to stick together unless immersed in water. If gill filaments are sticking together, they cannot operate efficiently. This is why a fish out of water will suffocate as it dries out even though it is surrounded by oxygen.

Modern amphibians get rid of carbon dioxide by diffusing it out through thin skin. Skin respiration is effective but has serious drawbacks. First, a large area of body surface must be devoted to gas exchange. Second, the skin must be thin and soft; scales, thick dermal layers, or other protective armor are out of the question. Thus, most skin-breathing amphibians rely on stealth, agility, camouflage, or poison defenses for protection. Also, since thin skins are susceptible to the desiccation problem, most amphibians are restricted to moist environments. And finally, considerations of the ratio of body surface area to body volume dictate that skin respiration will work only for relatively small animals with a sufficiently high surface area to volume ratio. Consequently, most living amphibians are rather small.

For larger animals both in and out of the sea, the gas exchange site for ridding one's body of carbon dioxide tends to be localized and richly supplied with blood vessels; both lungs and gills meet these specifications. For reptiles, birds, mammals, and even a few fish, the solution to the carbon dioxide problem is rapid breathing. A lung full of inhaled air exchanges CO_2 for O_2, refreshing the blood supply, before carbonic acid can build up to dangerous levels in tissues. Large volumes of air must be processed in this fashion to complete the CO_2–O_2 transfer because the gas exchange is only partial.

The development of lungs able to satisfy an animal's respiration requirements without help from skin or gills is a major evolutionary accomplishment. No longer must skin be devoted to respiration, and protective structures such as scales, hair, and feathers can safely be formed on the animal's skin surface. The use of lungs requires muscular pumping, which in mammals is accomplished by the rib cage and the diaphragm. Small amphibians take partial advantage of pump-enhanced breathing by inflating and deflating a throat pouch. Crocodiles compress and expand their lung volume by jerking their livers rapidly to and fro. Many terrestrial animal lineages probably skipped the skin-breathing phase altogether; the first land vertebrates were fairly large and probably had rib cage respiration or something akin to it.

Arthropods Arrive on Land

It seems likely that the first animals to solve the challenges of life on land were arthropods. It also seems likely that the animals' emergence onto land followed the appearance of plants, for the simple reason that plants could provide food for animals. This does not rule out, however, the possibility that the very earliest land animals made landfall before plants and that they fed on bacteria.

The soil habitat appears to have been a route onto land favored by different arthropod groups. Arthropods taking this route often underwent evolutionary miniaturization as an adaptation to life in the small spaces between soil particles. Small size may also have permitted a simple diffusion form of respiration. Such miniaturization may make these animals more difficult to classify, because some of the key morphological features that could be used to establish evolutionary links to larger animal ancestors in the sea have been lost to miniaturization.

Two related types of living arthropods will serve as the focus for our discussion of early land animals. One type is the lobopods, typified by the onychophorans or walking-worms and the second type, the myriapods, includes such familiar animals as the millipedes and centipedes. (Myriapod means "countless feet".)

A lobopod is a creature with a caterpillar-like appearance, bearing three or more pairs of unjointed legs. Lobopods either are arthropods or have a close evolutionary relationship to arthropods. Because so little is known about them, we believe that it would be premature to create for them a higher taxonomic category[1] (such as phylum). Onychophorans (figure 7.1) are extant members of the lobopod group. They look something like a centipede with swollen limbs. Only about eighty species are alive today, and they live primarily in the southern hemisphere in moist environments with wet leaves and rotting wood. In the winter, the Brazilian onychophoran *Peripatopsis acacioi* burrows as much as a meter into soil to avoid surface variations in temperature. The largest living onychophoran species, *Macroperipatus* from Trinidad, reaches 15 cm in length.

The legs, antennae, and trunks of onychophorans lack the true segmentation of arthropods but instead show ringlike annulations. Like arthropod legs, onychophoran legs are moved by internal musculature. But there the similarity ends. Onychophoran legs are wholly unlike the jointed legs of arthropods; the term "lobopod" refers to the lobelike shape of the legs. Many butterfly caterpillars have (in addition to their six adult-style jointed legs) rearward appendages somewhat like those of lobopods.

The unusual group of miniaturized animals called tardigrades or "water bears" are also lobopods. Tardigrades and onychophorans both have claws and spines at the ends of their walking or crawling legs; such features seem to be typical for the legs of lobopods.

Because they were soft-bodied, the ancient common ancestors

FIGURE 7.1
The living onychophoran or walking-worm *Euperipatoides kuczarti*. Length of animal is approximately 5 cm.

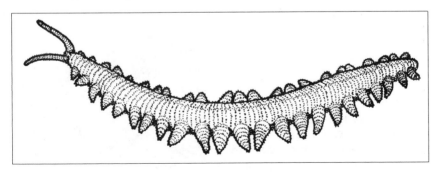

of onychophorans, tardigrades, and other "lobopodians" have a virtually nonexistent fossil record—nonexistent, that is, except in certain marine sediments of the early Cambrian. A peculiar body fossil was discovered in the 1920s in Germany and was claimed to be of Proterozoic age. It was later shown to be early Cambrian. The fossil, *Xenusion auerswaldae,* caused a minor sensation because it appeared to be the missing link between Cambrian animals and their mysterious pre-Cambrian past. *Xenusion* (figure 7.2) was placed in the taxonomic group Onychophora, a particularly satisfying assignment for scientists because it also seemed to provide an evolutionary link or intermediate between annelid worms and arthropods, two phyla with true segmentation. The onychophoran affinities of *Xenusion,* however, were challenged by some scientists (us included) who argued that this headless specimen may have been unrelated to onychophorans.

The *Xenusion* controversy took a sharp turn in 1989 with the discovery of a complete specimen of another enigmatic fossil, *Microdictyon.* Before this discovery, *Microdictyon* was known only from irregularly shaped, porous phosphatic skeletal plates recovered from limestones, which were usually obtained by dissolving the rock with acetic acid. These plates, once part of a multielement skeleton, are called sclerites. Figure 7.3 shows a typical, incomplete specimen of *Microdictyon,* visible on the naturally weathered surface of a piece of marine limestone.

Several Chinese scientists once classified *Microdictyon* as

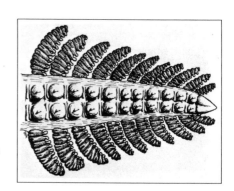

FIGURE 7.2

Scientific opinion has varied regarding the true nature of the Cambrian fossil *Xenusion,* but it is now clear that *Xenusion* is an onychophoran-like metazoan. Length of specimen is approximately 5 cm.

skeletons (more accurately, "tests") of radiolarians, unicellular marine protoctists. In 1989, however, other Chinese scientists found and described a complete specimen of *Microdictyon* from Chengjiang, China. It was an astonishing find, for it showed a fossil of a mostly soft-bodied lobopod with two rows of netlike plates or sclerites running down the length of its dorsal side. The fossil, *Microdictyon sinicum*, has been referred to as an armored lobopodian. *Microdictyon* sclerites (figure 7.3) are now interpreted as being directly homologous (i.e., derived from the same ancestral organ or body part) to a single spine in *Xenusion* (figure 7.2). *Xenusion* must now be interpreted, as it was originally interpreted, as a lobopod.[2]

Cambrian lobopods are related to the common ancestor of all arthropods—joint-legged organisms—both marine and terrestrial. The phylum Arthropoda includes all joint-legged organisms. One might think that myriapods, with their general resemblance to lobopods should be the most primitive of arthropods. Myriapods, however, are decidedly not the ancestral type.

The ancestral state in arthropods was probably that of polyramous appendages, in other words, legs with many branches. Some groups, during their evolutionary history, reduced the number of leg branches to just two, as for instance in the trilobites. Well preserved trilobite fossils indicate that these extinct arthropods had biramous appendages, with the upper branch typically bearing gills and the lower branch dedicated to locomotion.

FIGURE 7.3

Sclerite of *Microdictyon,* a Cambrian marine lobopod, on naturally weathered limestone surface. Though these sclerites are fairly common, not until 1989 did anyone know what the animal looked like. Greatest dimension of specimen is 1.3 mm.

An even more derived (evolutionarily modified) state in arthropods is reduction of the number of branches of each leg to just one. Arthropods with only one leg branch are referred to as being uniramous. Most modern arthropods are uniramous. The oldest arthropod with only one branch per leg is a Cambrian marine form called *Cambropodus gracilis*. Looking something like a smashed marine centipede, this form may have been ancestral to all subsequent uniramous arthropods on land such as insects and millipedes. The uniramous legs of *Cambropodus* are in contrast to the legs of forms like *Lepidocaris* (figure 6.57) and *Xenusion* (figure 7.2), which have polyramous, multibranched legs more akin to the ancestral arthropod condition.

Myriapods belong to the class Diplopoda, so named because they have two (uniramous) legs per body segment. Myriapods make an early and conspicuous appearance on land by virtue of their large size. In many parts of the world, modern millipedes can still reach lengths of ten centimeters or more.

First known from the Silurian, fossils of myriapod-like organisms are also found in the Rhynie Chert. A 370-million-year-old fossil centipede named *Devonobius delta* has been recently described from strata near Gilboa, New York.[3] Later occurring Paleozoic myriapods include larger forms such as the genus *Euphoberia* (figure 7.4) and *Acantherpestes major* (figure 7.5). *Euphoberia* and the giant *Acantherpestes* belong to an extinct Paleozoic class of myriapods called the archipol-

FIGURE 7.4

Euphoberia armigera?, a terrestrial, Coal Age myriapod. Legs are visible in the inner part of the body curve on the right side. Distance from tip to tail in curved position is 5 cm.

ypods. Some species of archipolypods may have been aquatic but most were probably terrestrial.

One group of myriapod-like organisms, the arthropleurids (named for their jointed dorsal skeleton), became even more conspicuous than the archipolypods during the Coal Age on account of their enormous size. The arthropleurids of the late Carboniferous were possibly descended from delicate ancestral forms seen in the Rhynie Chert. Early arthropleurids, some belonging to the genus *Eoarthropleura*, are reported from Devonian localities such as Ludford Land in Great Britain and Gilboa in New York State. By the late Carboniferous, the arthropleurid genus *Arthropleura* had attained lengths exceeding two meters. Tracheids and fragments of plant epidermis have been found in the guts of arthropleurids, indicating that these myriapod-like creatures ate plants. The flattened profile of arthropleurids probably allowed them to, in search of food, wedge their bodies into crevices between piles of decaying plant matter. This profile has earned arthropleurids, and other myriapods of similar shape, the nickname "litter splitters."

Earlier concerns that arthropleurids might have been aquatic were definitively settled by the discovery and description of fossil trackways thirty centimeters or more in width. These trackways were clearly formed on land. One such trackway records an animal that was weaving its way through a forest of *Calamites* that grew on a sheetflood deposit in an alluvial fan environment. Drier conditions prevailed in this ancient envi-

FIGURE 7.5

Fragment of *Acantherpestes major,* a long Carboniferous myriapod. Modern myriapods harbor unique Hypersea organisms within their guts. Maximum length of nodule is 18.1 cm.

ronment, confirming the terrestrial habitat interpretation for this arthropleurid.[4]

Both the myriapods and the lobopods included some armored forms. Armored lobopodian morphology of the early Cambrian is faithfully echoed much later in a Coal Age myriapod: the oniscomorph millipede fossil *Amynilyspes wortheni* from the Mazon Creek locality of northeastern Illinois. *Amynilyspes*, several centimeters in length, has a double row of spines (the dorso-lateral spines) along the dorsal side of its exoskeleton. These spines are reminiscent of the protective array of spines in the much older lobopod *Xenusion*. The myriapod was able to roll up, and the long spines likely served as a defensive mechanism against predators.[5] The long spines, however, would have precluded *Amynilyspes* from burrowing, though a burrowing lifestyle appears to have been common in other myriapods of this period.

Myriapods of the Paleozoic had just cause to develop heavy armor. In the early Devonian, the scorpion *Praearcturus* reached one meter in length. Considering its large size, with a stinger to match, *Praearcturus* was likely to have been the most fearsome, fully terrestrial predator of its day.[6]

Beerbower's Litter Box Hypothesis

In our first book, *The Emergence of Animals: The Cambrian Breakthrough*, we argued that the heterotrophic, predator-rich ecosystems of the Cambrian sea were transposed onto land when animals made the transition to terrestrial habitats. Indeed, the fossilized communities of early land animals are rich with arthropod predators—so rich that many of the best preserved components of the fauna are its predators. The jaws of a modern centipede (figure 7.6) and an amber-trapped Miocene jumping spider (figure 7.7) provide a visual reminder of how long these predatory life modes have been important on land.

Richard Beerbower makes the interesting argument that

with the rise to dominance of vascular plants in the Paleozoic, the usual direct link between primary producers and consumers may have temporarily disappeared. According to his hypothesis, there were very limited numbers of herbivorous animals on land until the early Carboniferous. Beerbower thinks that the reason for his hypothesized lack of early land herbivores lies in the structural compound lignin, which is abundant nowhere except in vascular plants. The synthesis of lignin yields several toxic chemical byproducts. Plants are unable to excrete these toxic chemicals, so they store them instead in their cell walls and dead tissues. Here they provide service to the plant as deterrents to herbivores. Such defenses also afford modern trees some protection from heartwood rot.

In effect, lignin-forming plants were incidentally defended against grazing on account of the toxicity of the secondary products of lignin synthesis. Furthermore, plant tissues are nutrient-poor, and are especially low in sodium and aromatic amino acids. The complex carbohydrate polymers making up

FIGURE 7.6
Fangs of a modern centipede from Kyoto, Japan. Width of specimen in view is 6 mm.

FIGURE 7.7
Jumping spider (genus *Corythalia*) in amber from the Dominican Republic, identified by its smooth bullet-shaped pedipalps. Dominican amber is at least early Miocene in age (approximately 23 million years old). Note also the chalcid wasp above the spider. Perhaps this particular spider was trying to catch that wasp when it became mired in the sap death trap? Length of spider is 3.8 mm.

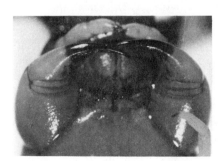

cell walls are difficult or impossible for animals to digest, and animals must rely on saprophytic fungi (or in some cases the animals' symbiotic gut protoctists) to break down leaf litter components and improve their nutritional qualities. Only in these ways could land animals tap into the land food chain. The early arthropods that colonized the land must, therefore, have initially been detritivores, not herbivores. Beerbower's idea regarding the absence of herbivores before the late Paleozoic has been called the Litter Box hypothesis. In this view, early land faunas consisted of animals incapable of digesting toxic substances accumulated in the tissues of living plants. Rather, they browsed the decomposing plant litter of the soil, depending on fungal members of the community to first make plant products palatable.[7]

The contention of the Litter Box hypothesis that land herbivory was rare or absent in the Devonian is attractive to us. Indeed, it has similarities to the Garden of Ediacara hypothesis. Both ideas postulate that important ecological life strategies in the sea and on land were delayed in their development. Put differently, both the "garden" and the "litter box" owe to lag times between an evolutionary breakthrough (large marine creatures in the one and vascular land plants in the other) and the full exploitation of these new developments by other organisms (large marine predators and terrestrial herbivores, respectively).

Before the Litter Box hypothesis can be accepted, an adequate explanation will be needed for the apparent wound tissue seen in specimens of two species of early Devonian vascular plants: *Psilophyton dawsonii* and *Psilophyton coniculum*.[8] In the latter species, in particular, a large number of wounds are found along the lengths of some of the fossil stems. Although the cause of the wounding to *Psilophyton* is not evident, it is quite possible that animal herbivores were responsible. Furthermore, the unicellular spinose hairs of *Trichopherophyton*, the newly discovered zosterophyll from the Rhynie Chert, can be interpreted as an antiherbivore adaptation.

Harlan P. Banks has recently argued that fossilized wounds on early Devonian *Psilophyton* branchlets are suggestive of herbivory by arthropods equipped with chewing and sucking mouthparts. He sees evidence for multiplication of cells beneath the wound, enlargement and in some cases explosion (lysing) of the damaged cells, and formation of protective tissue or periderm. According to Banks, these features indicate that the plants were alive when injured; these wounds are not to be explained by the munching of detritivores after the plant had joined the soil litter.

Unexplained wounds that show evidence of healing are also seen in the Rhynie plants. In defense of the Litter Box hypothesis, however, William A. Shear attributes these injuries to burns from volcanic cinders or other mechanical damage.[9] We find Shear's volcanic cinder suggestion to be improbable, but this does not necessarily mean that the damage was due to arthropod herbivores. In further defense of the Litter Box hypothesis, Paul A. Seldon argues (incorrectly, in our opinion) that spores and intact spore masses in early arthropod droppings (coprolites) were accidentally ingested by detritus-eating arthropods.[10] We believe it likely that at least some of the early land animals were seeking out and eating spores—and that their digestive systems were therefore already capable of penetrating the tough spore coats to extract nutrients from within.

Andrew J. Jeram claims that the low diversity of Paleozoic scorpions supports the Litter Box hypothesis. His argument is that scorpions could not evolve into very many types because all they had to eat were a few kinds of generalized detritivores—litter-eating arthropods. There was no wonderful array of different types of arthropods specialized for feeding on living plants. This argument is interesting but fails to take into account the fact that numerous other predators besides scorpions existed at the time, such as the trigonotarbids (relatives of spiders; figure 6.58) and centipedes, including an entire order of centipedes no longer living (represented by *Devonobius delta*).

If a number of Devonian herbivores are discovered, the Litter Box hypothesis must head for the waste basket.

In an apparent departure from the Litter Box hypothesis, Beerbower has recently argued that the reason sporangia formed on the tips of erect stems of early plants (figure 6.13) was not only to enhance spore dispersal but to protect them from animals capable of seeking out and eating the nutritious spores.[11] He posits that erect sporophyte stems with terminal sporangia (sporangia at the tips) were an adaptation to minimize losses to small herbivores and spore eaters largely confined to the soil surface. The early appearance of terminal sporangia in the fossil record thus may be indicative of the evolution of the first terrestrial herbivores that posed a real threat.

In any case, and whenever and however a herbivorous group evolved, early animal herbivores surely concentrated (as do human vegetable lovers today) on eating ephemeral plant parts such as spores, ovules, seeds, and young leaves before these parts have accumulated significant concentrations of toxins. Some modern insects and vertebrates, of course, have enzyme complexes to detoxify plant chemicals, and some animals store ingested plant toxins for use in their own defense. But most animals that eat plant matter are still heavily reliant on members of other kingdoms to detoxify their food. Termites would starve without their Hypersea complement of wood-digesting bacteria. Several species of bark- and wood-boring beetles feed primarily on yeastlike forms of fungi that find beetle tunnels ideal for their own needs. Some of these beetle species have evolved mycangia, specialized pouches for carrying their favored food fungus. The mycangia inoculate pristine wood as the beetles crawl through it, thus "seeding" a new food resource for the beetles, who later return to graze on the fungus. Other insects possess mycetomes, specialized organs in which fungi are permanently stored, perhaps because they provide vital organic compounds to the insect.

A Hypersea Alternative to the Litter Box

Arthropods clearly need microbial help to digest plant matter. This help can occur outside the animal's body as per the Litter Box hypothesis—or the help can occur inside the animal's gut, in the realm of Hypersea. As noted earlier, arthropleurids from the late Paleozoic have been found with tracheids and plant cuticle in their guts, indicating a plant diet. A six-foot-long Coal Age arthropleurid almost certainly had fungi and other microbes within its gut to help it digest plant matter.

Indeed, a large living millipede is a virtual walking ecosystem of gut-associated organisms. Attached to the chitinous gut lining of living millipedes are filamentous actinomycete bacteria, protoctists, and even nematodes—all infesting the myriapod's body while it thrives in perfect health. Furthermore, there is a major class of fungi, the Trichomycetes ("hair fungi") which are found *only* within the guts of these myriapods and in certain other organisms such as centipedes. Trichomycetes of the genus *Enterobryus* attach to the millipede gut lining by means of a suckerlike holdfast. The holdfast itself is often associated with radiating clusters of filamentous actinomycete bacteria. It is fair to speculate that the giant arthropleurid myriapods of the late Paleozoic may have had, in addition to trichomycetes, an even more complex and interesting gut biota because their guts were so large. Unique groups of fungi and other gut organisms may have perished when arthropleurids went extinct. We may never know for certain, however, because arthropleurids died out in the Paleozoic, and we expect it may be a long time before anyone discovers a well-preserved fossil record of arthropleurid microbial gut contents.

Suffice it to say that biotas associated with the arthropod gut are, and have been for quite some time, a major manifestation of Hypersea. Organisms such as millipedes and termites have large numbers of and many different kinds of organisms on the insides of their bodies. Some of these forms are found nowhere else. Such a level of gut diversity is unknown in the sea, even among marine animals living in or subsisting on hard-to-digest driftwood. The arthropod gut is therefore one of the most pronounced examples of the key characteristic of Hypersea: the nestedness of land life.

Two possibilities exist for the origin of trichomycetes fungi and their habitat restriction to the arthropod gut. First is that they may represent a formerly widespread marine group that reached land by colonizing the guts of myriapods, and whose free-living forms subsequently went extinct in the ocean or in freshwater aquatic environments. In this case, the myriapod gut acts as a fungal refugium. Alternatively, trichomycetes may have evolved from other fungi into an entirely new class of fungi, and the evolution of this major new group occurred entirely inside arthropods.

Either way, as refugium or evolutionary incubator, we see how Hypersea manifest in the arthropod gut can retain or create, respectively, a level of diversity higher than that present in the oceans.

Arthropods and arthropodlike animals not only serve as refugia or evolutionary loci for smaller organisms; these creatures also become beneficiaries of the hypermarine diversity-enhancing effect. This gives us reason to believe that Hypersea is a new and general scientific principle. That is, Hypersea is a previously unrecognized biogeophysical entity—the body fluids of eukaryotic organisms and the inhabitants of those fluids—that allows and encourages a tremendous amount of diversity of organisms by creating untold numbers of isolated living environments.

The Parasitic Pentastomes

Consider the case of pentastomes. Pentastomes are unique chitinous, worm-shaped creatures that merit their own phylum; they are all internal parasites of land vertebrates (figure 7.8). They attach to the host tissue by a row of hooks on their heads and feed on the blood and tissue fluids in the lungs or air passages of the host. Pentastomes can attain sizes of up to 16 cm. They have a straight gut and an amply sized reproductive system, a gonadal investment which is typical of endoparasites. A pair of large glands apparently produces secretions to protect the pentastome from immunological defenses of the host.

Pentastomes typically have two hosts during their life cycles. Male and female pentastomes mate within the final (or definitive) host. Their eggs pass out through the host's saliva, mucus secretions, and feces; the eggs are eaten by intermediate hosts, which can be fish, amphibians, small reptiles, small mammals, or insects. Inside their intermediate hosts, the pentastome eggs hatch into four- to six-legged larvae. The larval form bores through the gut of the intermediate host and enters vital organs, where it then feeds and grows. When the intermediate hosts (possibly weakened by the infection of the parasite) are captured and eaten by a predator, such as a snake, the predator becomes the final host. The young pentastomes attach to the nasal passages and lungs of the predator and complete their life cycle. Fortunately for our species, though pentastome juveniles can infect humans, they do not survive to maturity.

The approximately 90 extant species of pentastomes are most common as parasites of reptiles, but they also occur in birds

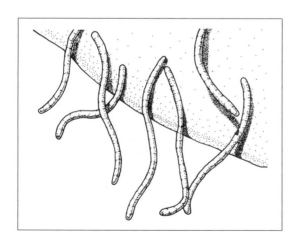

FIGURE 7.8

Population of *Kiricephalus pattoni,* a snake parasite from Taiwan. This pentastome is endoparasitic on the blood and tissue fluids of a snake; it attaches itself by a row of hooks on its head to the lining of the snake's lung. The parasite bodies are up to 16 cm long.

and mammals. In spite of their wormlike shape in one phase of the life cycle, pentastomes periodically molt their chitinous cuticle. This fact and other evidence, such as larval legs, suggests that pentastomes are actually closely related to arthropods. Virtually all arthropods, and onychophorans as well, shed their chitinous exoskeleton as growth proceeds. This suggests evolutionary ties to a common lobopod ancestor also capable of molting.

In the case of pentastomes, evidence can be mustered in support of both the "refugium" hypothesis and the "new evolutionary locus" hypothesis for their origin. In support of the latter, and based on similarities in larval morphology and development, it has been proposed that pentastomes are direct descendants of branchiurids or fish lice.[12] Branchiurids occur in both marine and fresh water; they attach themselves to their fish and tadpole hosts by means of suckers and hooks. Figure 7.9 shows the modern branchiurid *Argulus*. There are about 130 species of extant argulids. Note the biramous appendage— thought to be a more primitive condition than that of arthro-

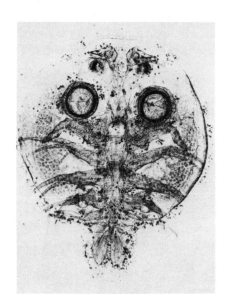

FIGURE 7.9

Branchiurid *Argulus*. This arthropod, commonly known as a fish louse, is parasitic on fish skin. No branchiurid fossils have ever been confirmed, but certain primitive characteristics, such as the biramous (two-branched) legs seen here, make it seem likely that the group has its origins in the Paleozoic. Length of specimen is 4.7 mm.

pods with uniramous legs. Figure 7.10 is a detail view of *Argulus'* suckers, small compound eyes, and hooks.

Branchiurids are unknown as fossils. However, their seemingly primitive characteristics make it quite likely that they have an ancestry going back to the Paleozoic. The Carboniferous fossil *Cyclus americanus* (figure 7.11) has been compared to branchiurids, but the preservation of known fossils is not detailed enough to prove or disprove this suggestion. All that can be said with confidence about the fossil is that *Cyclus* is an

FIGURE 7.10
Enlargement of figure 7.9, showing suckers, compound eyes, and hooks. Branchiurids and the pentastomes of figure 7.8 may be closely related. Distance between outer edges of suckers is 2.2 mm.

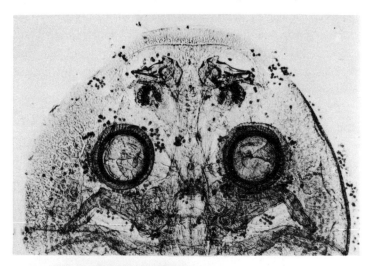

FIGURE 7.11
Cyclus americanus. This Pennsylvanian fossil may or may not be a branchiurid. The animal's mouth is toward the bottom of the figure; dark trace of gut is visible. Width of rock nodule is 22 mm.

animal with a gut, because the fossil gut trace makes a distinct stain in the fossil.

Fossil evidence in the form of a recent discovery can also be presented in support of the refugium hypothesis for the origin of pentastomes. Fossils of a soft-bodied organism, exquisitely preserved by postmortem phosphatic mineralization, have been described by Dietmar Andres as Paleozoic marine pentastomes (figure 7.12). If correctly interpreted, these would not only be the first examples of fossil pentastomes; they would also place the origin of pentastomes back to the early Paleozoic, as the fossils are Ordovician in age.[13]

As in the case of trichomycetes, two options exist for the evolutionary history of pentastomes. Either they evolved from branchiurids after somehow gaining access to the tissues of land vertebrates, or pentastomes were an originally free-living marine group that infested land animals *as pentastomes*—and subsequently went extinct in the seas. More fossil data and perhaps gene sequence comparisons between branchiurids and pentastomes may help decide the question.

The first possibility, that the marine precursors of pentastomes evolved into pentastomes only after migrating to Hypersea, is plausible because ingestion of branchiurids by fish-eating reptiles could give the putative branchiurid ancestors of pentastomes easy access to the interior of reptiles. In this way the parasites have the opportunity to associate with the larger animal, now as an internal parasite (endoparasite) rather than as a pest on the outside. Assuming that the ingested branchi-

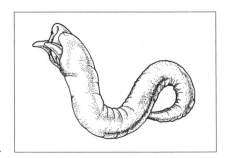

FIGURE 7.12
Reconstruction of a fossil animal from Ordovician strata of Sweden. This form has been interpreted as an early, free-living pentastome. Length of specimen (if straightened out) is approximately 1.4 mm.

urids could avoid being digested, they could evolve a parasitic lifestyle inside the new host and become the first pentastomes.

The alternate or refugium hypothesis is probably more likely, however, assuming that the Ordovician phosphatized fossils of figure 7.12 are truly pentastomes. If pentastomes already existed in the sea before entering land animals, then the marine pentastomes would have to be ancestral to the land pentastomes, and the reptilian parasites are the last remnants of a once free-living phylum. Either way, whether branchiurids gained a new habitat for evolution within the bodies of land vertebrates or whether pentastomes found a refuge from marine extinction within the bodies of land vertebrates, the body fluids of land vertebrates ended up serving as an evolutionarily important reservoir of Hypersea. As is true of virtually all such parts of Hypersea, this reservoir makes a significant contribution to the total species diversity of organisms on land.

EIGHT

Parasitism in Hypersea

Although intrinsically repulsive to humans, parasites are fascinating organisms with increasingly appreciated importance for evolutionary theory. The specificity of parasites to their hosts can be astonishing. The "chewing lice" of the insect order Mallophaga are so host-specific that each species of bird or mammal is host to one or more *unique* species of lice. Each species of Mallophoga can, in a way, be considered endemic to the habitat afforded by one species of bird. This host specificity is so strong that ornithologists occasionally use the insects as auxiliary traits in the description of new bird species. One can imagine such host specificity making significant contributions to the total diversity of the terrestrial biota.[1]

The influence of parasites on their host species may be extremely important. Notably, protection from parasites may be a major reason that sexual reproduction has remained common among plants and animals. The main question regarding sex is, why should any lineage bother to maintain two sexes when (1) asexual reproduction is a much more efficient and less risky process and (2) from a female point of view, males are akin to parasites, because a female can pass on only half her genes to each of her progeny. It seems that there should be strong selective pressures to eliminate sex, but in most animal groups this has not happened. Why not?

Parasites may be the answer. Sexual reproduction can maintain resistance traits in a host population. These traits may have temporarily lost their usefulness but could prove vital to survival in the future. This crucial advantage is not shared by the cloned populations derived from asexual reproduction. Furthermore, cultures of fungal clones seem to need periodic sexual recombination or they lapse into senescence. It has even been argued that death itself is a method of protecting one's relatives and progeny from parasites. Death of the host often means death of the parasite. William Hamilton has gone so far as to suggest that sex itself originated because it results in genetic recombination, a reshuffling that may help organisms fend off parasites. In support of this hypothesis, Curtis Lively has reported a New Zealand snail species that undergoes asexual reproduction where parasites are scarce, but sexual reproduction in habitats where its parasites are common.[2]

Navigating Through Hypersea

Some parasites are capable of altering the actual behavior of their animal hosts. An oft-cited example of this is the parasitic worm *Leucochloridium paradoxum* and its effect on its host, pulmonate land snails of the genus *Succinea*. *Succinea*, like many land snails, normally prefers low levels of light. But when parasitized by *Leucochloridium*, the snails become attracted to higher levels of light. The parasites mature inside succineid snails and then invade the snail's eyestalks in large numbers and make the eyestalks brightly colored. The swollen, throbbing, colorful eyestalks bear a striking resemblance to caterpillars or other tasty (from a bird's perspective) insect morsels. By interfering with the snail's optic nerves, the parasites cause the snails to seek higher light intensities, for instance, near the tops of the plants on which the snails feed. When these conspicuous eyestalks reach the tops of the plants, they are spied by birds who peck them off. These birds are the next host in the *Leucochloridium* life cycle. *Leucochloridium* thus deftly solves for itself the transportation problem from one host

to the next; this parasite's life cycle can be thought of as
employing an effective navigational trick to pass from one
reservoir of Hypersea to the next.[3]

A parasitic acanthocephalan (the spiny-headed worm *Plagi-
orhynchus cylindraceus*) induces a similar behavioral change in
its host, the common pill bug (the isopod crustacean *Armadil-
lidium vulgare*). Parasitized pill bugs develop "wanderlust" and
crawl around out in the open, where they can be spotted
and swallowed by starlings: the next Hypersea reservoir to be
colonized by the parasite. Janice Moore, who has studied this
relationship, notes that the isopod is in effect a parasite "dressed
up" to look like a pill bug. Moore feels that, since such host-
parasite relationships are difficult for biologists to spot without
detailed study, they may be much more common than anyone
previously realized.[4]

A related, but more complex, evolutionary strategy is employed by the sheep liver trematode *Dicrocoelium dendriticum*.
This parasite must pass through a snail and then an ant before
reaching the sheep, its definitive host. Since ants are not an
intended part of the sheep's diet, it is something of a problem
for *D. dendriticum* to make the leap from ant to sheep. This
problem is solved by inducing fairly intricate behavioral
changes in the ants.

Normal ant behavior is to be active on the surface of the soil
during much of the day but to retreat to burrows during the
midday heat of summer. Infected ants behave much differently. They climb to the tip of grass blades in the morning and
clamp their jaws to the tops of grass blades, thus greatly
increasing the possibility that sheep will accidentally swallow
them along with the grass. Like their healthy comrades, the
parasitized ants do retreat down the blade to escape the potentially lethal noon sun, but they then return to the blade tops in
the cooler afternoon. The behavior is repeated each day until
the ant either dies or is eaten by a sheep.[5]

In addition to maintaining sex and altering host behavior,
parasitism has apparently played a key role in bringing many

types of aquatic organisms into the terrestrial biota that might otherwise have been denied access because of various biological and physical barriers. Parasitism may have been the only method by which fungi were able to leave aquatic habitats and begin fully subaerial life. Since fungi are completely dependent on other organisms to create their food, it seems reasonable to assume that the first land fungi (which unlike animals are not motile) must have been closely associated with a food-source organism such as an early plant. The fossil record of the Rhynie Chert, and the robust size of early *Baragwanathia*, suggest that such association was of a hypermarine character at the outset.

Lichens are as close as fungi get to living an autotrophic existence. Lichens may be considered an example of controlled parasitism, and laboratory studies demonstrate that such parasitism has been modified by resistant phycobionts (algae) able to survive enslavement by the mycobiont (fungus).[6]

Animals also used parasitism as a route to land, although they probably have not used it as a "stepping stone" to a free-living terrestrial life in the same way hypothesized for fungi. Nevertheless, a diverse group of higher taxonomic categories of animals are represented on land only by parasitic forms. One of the most interesting cases of terrestrial animal parasitism involves the example of the pentastomes discussed in chapter 7.

Considering the importance that dinosaurs played in Mesozoic terrestrial ecosystems, we think that it is safe to infer that pentastomes (which favor dinosaurs' close relatives, the reptiles and birds, as hosts) did very well, in an evolutionary sense, during the Mesozoic. We think it is entirely reasonable to expect that pentastomes parasitized dinosaurs, especially forms such as the Cretaceous dinosaur *Baryonyx walkeri* (figure 8.1; the genus name means heavy claw), which was a probable fish-eater.[7] By ingesting but not digesting a pentastome ancestor attached to the skin of its fishy prey, *Baryonyx* might have opened the door into hypermarine habitats for pentastomes—if they had not entered Hypersea already.

Pentastomes may have been the largest land parasites of all time. If modern forms can reach 16 cm in length, it is not inconceivable that a large *Brachiosaurus* dinosaur could have harbored pentastome parasites several meters or more in length.

The blood and lymph of the intermediate and final hosts, from the point of view of a pentastome, represents a dispersed ecosystem, to be colonized by taking advantage of those brief instants when one oasis of Hypersea comes in contact with another—as during predation. Hypersea is a new type of ecosystem, and pentastomes are either an animal phylum that evolved within this new ecosystem or a group that originated in the oceans, spread to Hypersea, and then went extinct in their ancestral home. Like a string of kettle lakes connected by streams only during times of flood, the bodies of land animals are lakes within the Hypersea hydrosystem—connected by the fluids they exchange during moments of contact.

Parasites are so restricted by their preferences for particular host types, host organs, and even host tissues that it may at first

FIGURE 8.1

Claw of *Baryonyx walkeri*, a recently discovered dinosaur that may have been host to pentastomes. This extremely large claw is presumed to be for catching fish. Claw is nearly 25 cm long. Scale bar = 5 cm.

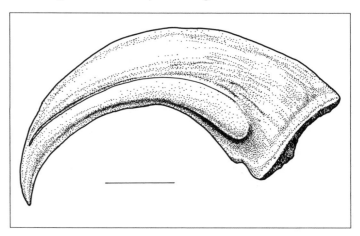

seem implausible to consider host tissues as a kind of aquatic environment. It would be analogous to a group of marine organisms being restricted to a particular rock on the sea floor, which appears to be a rare situation in the sea if it occurs at all. But Hypersea is not like open ocean. It is more like a series of bays, estuaries, and inlets, or a dispersed chain of coral reefs, each with its own endemic set of species.

Parasites do have fairly specific "itineraries" as they pass through their hosts and the organs of their hosts. But these movements are comparable to the migrations of anadromous fish from fresh water to the sea and later back to a particular— not just any—freshwater stream. Sometimes, of course, migration goes awry; a fish will sometimes return to a stream other than that where it was spawned.

Consider the tragic case of three-year-old Roberta Hanson. Roberta was infected by the common parasitic roundworm *Ascaris lumbricoides*, probably while she was playing in her backyard sandbox. *Ascaris* larvae usually travel to an animal's lungs and liver, then on to the epiglottis for a ride back down the esophagus and out the length of the alimentary canal. Worm eggs are dispersed via the infected animal's feces. In Roberta's case, however, the worm got lost on its migratory route and ended up in her brain, carrying with it a load of fecal bacteria. These bacteria are usually beneficial when they reside in a human's gut, but in the brain the microbes were deadly. The girl's brain formed a huge abscess. Within a year she was dead; an autopsy showed that disease had caused her entire brain to degenerate. This condition was probably already incurable by the time Roberta began to show symptoms.[8]

Although this tragic case was certainly no help to the parasite either, it is easy to imagine parasitic wanderings that actually prove beneficial to the parasite (recall one of the hypotheses of the origin of land pentastomes) by leading to new parasitic habitats and lifestyles.

Escaping the Trap of Host Extinctions

Hypersea shrinks and expands the most during and right after times of mass extinction. The mass extinction that marks the Cretaceous-Tertiary boundary, with its complete loss of the great dinosaurian groups, was likely to have been a time of extinction for many kinds of pentastomes as well. This raises an important question concerning migrations in Hypersea: how can a parasite survive the extinction of its only host?

Study of this problem focuses on monospecific host-parasite associations. No doubt there are good reasons why some parasites are restricted to a single host species. Perhaps there are advantages to the parasite in being so specialized, or perhaps it happens because of geographic or host-related isolation. In spite of any possible advantages, there are great risks for the parasite species because, if a parasite is restricted to a single type of host (i.e., if it restricts itself to only a single type of Hypersea reservoir), extinction of the host species will always mean extinction of the parasite species as well.

A possible escape from this type of evolutionary trap was proposed in 1990.[9] Imagine a parasite (like a pentastome) with diverse intermediate hosts (such as lizards, mice, cockroaches) but a preferred final or definitive host (a predatory reptile). Then, imagine that an aberrant form of the parasite in question restricts itself to the intermediate or secondary host, avoids going on to the final host for one reason or another, but is still able to reproduce. This would be an example of evolutionary retention of a juvenile characteristic. If this change were to occur during a time of mass extinction of final host forms, the juvenile-like (paedomorphic) mutant parasites would still be alive after the extinction event. The life cycle of the surviving parasites might then be "fleshed out" at a later time and a new final host could be parasitized. In terms of Hypersea, then, the intermediate hosts act as refugia for the parasite lineage in question, in the same way that the shallows surrounding isolated islands can act as refugia for marine organisms during times of extinction on the continental shelves. Interesting evolutionary bottlenecks such as these surely play a crucial role in the evolution of parasites.

Plant parasites could run into similar crises, and find similar opportunities, during times of general extinction. A blister rust fungus of the white pine has gooseberry or currant bushes as alternate hosts, another rust alternates between cedar trees and apple trees, and the black stem rust fungus of wheat has wild barberry bushes as an alternate host. If one of the hosts in the alternation were to go extinct, the parasitic fungus would be forced to temporarily restrict itself to a single host or go extinct as well. Indeed, foresters often control alternate host parasitic fungi by destroying one of hosts, generally the less valuable of the two.

Recall the acrimonious fungus (*Acremonium coenophialum*), which infects stems and seeds of tall fescue. The grass experiences several significant benefits as a result of fungus infestation. We speculate that this grass, a cool-season European perennial import, has had a long history (perhaps millions of years) of association with acrimonious fungus or a similar endophytic, symbiotic fungus. Fescue welcomes the association, but infection is not always present; there are circumstances when the fungus is not typically found within the fescue grass. Like an animal parasite in the midst of avoiding extinction while its definitive host goes extinct, the fescue symbiont may temporarily prefer other hosts and only at a later time return to the particular Hypersea inlet of *Festuca arudinacea*.

Hypersea, the Biogeophysical Entity

This is an appropriate moment to pause and review the definition of Hypersea. The term *Hypersea* as we have been using it in this book sometimes refers to a physical entity on the earth's surface. Specifically, Hypersea is the sum total of the bodies and body fluids of plants, animals, fungi, and protoctists that reside in terrestrial habitats, along with their associated parasites and other symbionts; it does not include free-living land bacteria lacking symbionts.

We regard this aggregation as special, as worth reifying, because by focusing on the evolutionary innovation of directed flow of nutrients and the colonization (and navigation) of a kind of "inner sea" one can extract new insights about terrestrial evolution. This leads to our other usages of Hypersea: Hypersea is also a new view of terrestrial evolution. The theory aspect of Hypersea will be the subject that closes this chapter. For now, we will review what we have already presented about Hypersea, the entity.

We have presented examples of Hypersea, manifest as the bodies of land organisms such as fungi, millipedes, reptiles, and, most importantly, vascular plants. Hypersea as the tissues of these multicellular land organisms of course comprises myriad uncountable separate parts, and many of the parasitic organisms that make Hypersea their home are quite adept at navigating from one reservoir of Hypersea (host) to the next.

Some microbes are capable of creating a biochemical camouflage to shield themselves from the immune systems of their hosts. Such camouflage is comparable to the visual camouflage of free-living organisms, for example, insects disguised as thorns or leaves to avoid predation. Metazoan parasites have, in addition to biochemical camouflaging, evolved sense organs that do not exist in their free-living relatives. These unique sense organs allow the parasite to navigate within the host. Schistosome flukes have several of these unique sense organs, including tentaclelike protuberances used for navigation by chemodetection.

Elements of Hypersea vary with regard to their geologic age. Some Hypersea organisms are geologically recent. Orchids and other epiphytic plants that send roots not into the soil but rather into the air or into the tissues of other vascular plants have undergone an astonishing diversification in the last few tens of millions of years.[10] Little fossil evidence exists for epiphytes before the Cenozoic, probably because the Hypersea reservoir required by epiphytes like orchids and mistletoe did not become extensive until the end of the Mesozoic, when magnolias and other angiosperm trees began to form dense forest canopies. Near the Cretaceous-Tertiary boundary, forests composed of multilayered canopies were fostered by a trend toward a more humid global climate.

Parasitism and hyperparasitism of eggs and larvae by insects such as the mymarid chalcid wasp (figure 8.2) and eulophid chalcid wasp (figure 8.3), quintessential inhabitants of Hypersea, had to await both the advent of flight in insects (so the adult parasitic wasps could seek out hosts) and the expansion of insects and other host arthropods into many different lifestyles. For insects, most of this expansion occurred during the late Paleozoic and into the Mesozoic.[11]

It appears, however, that many of the most basic Hypersea interactions are very ancient, dating back to or close to the origin of the earliest vascular plants. Recall our earlier analyses of the origin of mycorrhizae (chapters 1 and 6). Leaf spot

FIGURE 8.2

Fairy fly (mymarid chalcid wasp) in Dominican amber. This parasitic wasp is Miocene in age and is a mere 700 microns (seven tenths of a millimeter) in length. Mymarids live both as parasites and as hyperparasites. Spherical object is a bubble in the amber.

disease is a common pest of modern tropical *Hymenaea* trees. *Hymenaea* leaves trapped in Miocene amber (approximately 23 million years old) sometimes show evidence of infestation by leaf-spotting fungi (figure 8.4). These fungi live inside the living leaf, and they form their spherical reproductive structures there as part of the intra-leaf infestation.[12] Could these fungi be evolutionarily en route to adding another, above-ground "mycobiont" to the plant association, in addition to the mycorrhial fungi occurring below ground?

In summary, Hypersea as a biogeophysical entity is composed of plants, fungi, protoctists, animals, and their parasites, hyperparasites, and other symbionts. It is distinguished by biotic mediation and enhancement of nutrient flow. Whereas fundamental physical processes (river runoff, ocean currents, ocean upwelling, and evaporation) direct nutrient flow in the sea, Hypersea works nutrient flow by biotic innovations. Hypersea is continuously gaining and losing water through the biotically mediated actions of roots, leaves, lungs, and cuticle.

Hypersea, the Theory

In order to merit consideration as a scientific theory, a new idea must not only generate a good amount of corroborative evidence; it must be stated as a testable hypothesis. Out of the idea

FIGURE 8.3
Chalcid wasp in Dominican amber. This parasitic wasp is also visible in the upper edge of figure 7.7. It probably belongs to the family Eulophidae. Length of wasp is 0.9 mm.

must flow specific predictions which, if proved wrong, would "falsify" the stated hypothesis or hypotheses. These predictions must not be assembled post hoc; they must be made before new data is gathered or existing data is analyzed in a particular fashion suggested by the hypothesis. Too slavish an adherence to the doctrine of falsifiability and testability can blind one to the more intuitive (and often ultimately more successful) approaches to scientific work, but the "bottom line" for any scientific theory or hypothesis is that, at least in principle, it must be subject to test and ensuing verification or falsification. Otherwise it is not a scientific theory. Thus, if a hypothesis passes the test of new data or new analysis it still, of course, faces the hurdle of attracting sufficient interest among scientists to be incorporated into the day-to-day workings and thought processes of an identifiable research community. In short, it must be of use; it must be fruitful.

Thus far, we have surveyed the evolution of terrestrial plants, fungi, animals, and their symbionts and parasites from the

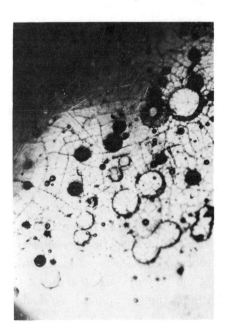

FIGURE 8.4

Fossil leaf in amber. The leaf was infected with leaf-spotting fungus. Such fungi are common parasites on the leaves of tropical plants today. The fungus is unidentified but we believe it to be an ascomycete, in which case the spherical structures are ascocarps. The largest spherical structure is approximately 700 microns in diameter.

standpoint of a Hypersea perspective. That is, we have pointed out specific evidence that corroborates our view that the biogeophysical entity we call Hypersea has made significant alternations to the physical and geochemical landscape. But we have not yet fleshed out predictions—significant predictions at risk of falsification, which when tested will either falsify or further corroborate Hypersea theory.

We believe that our Hypersea concept is a scientific outlook that can generate a family of testable hypotheses which, without this new outlook, might never be considered. These hypotheses will have important theoretical implications. We have already presented three such hypotheses in this book. First is that hyperparasites are more common on land than in the sea. Second is that, for at least the past 350 million years, the land biota has had several more higher levels of hyperparasitism than the marine biota has had. The third hypothesis is that those land plants able to make symbiotic associations with a variety of species will be more widely distributed than will similar species with less symbiotic flexibility. Following are four more testable predictions of the Hypersea theory. To our knowledge, none have yet been tested.

1. *Terrestrial biotas should have a higher total number of intermediate hosts or "vectors" in their parasites' life cycles than marine biotas have.* If host intermediates are just as common in marine environments as on land, after standardizing for total number of species, then Hypersea theory loses a good deal of its credibility.

This hypothesis should lend itself to straightforward and rigorous testing, but it will require biologists to devote more attention to looking for nested parasitism in the marine biota. In our view, this particular prediction is thrown out onto a level playing field; there is no unfair bias in favor of the Hypersea hypothesis, for many bizarre marine parasites are known to exist. Parasitic copepod crustaceans on marine fish attain such conspicuous sizes and bizarre shapes that they have been noticed since classical times (first recognized, no doubt, by early fishermen). Free-living copepods were not recognized at all until the seventeenth century. Members of the marine copepod genus *Xenocoeloma*, parasitic on marine polychaete worms, are so modified for parasitic life that adults are reduced to a mass of tissue in the host worm. *Xenocoeloma* can be recognized as a cyclopoid copepod only because of the presence of egg sacs and copepod larval stages.[13] Marine crustaceans themselves are infested by the poorly known funguslike protoctists called ellobiopsids, which form spore clusters on the outside of the animal's body.[14] But bizarre morphology is not the measure of this particular prediction—the number of intermediate hosts (if any) is what counts.[15]

Consider, too, the disparity in the length of the food chain between the sea and the land. The length of a food chain is measured by the number of steps taken in proceeding, say, from a primary producer like a plant to the herbivore to the carnivore on up to the top carnivore. The average length of a food chain in the sea is generally higher than the average length of a food chain on land, so there is no bias for the average land food chain to have greater opportunities for host intermediates because of greater length; in fact the opposite would hold. For example, mean food chain length in the Malaysian rain forest is 1.88, whereas mean food chain length in the surface zone of tropical seas is 5.92.[16] Also, marine ecosystems are older and would have had more time to develop linkages between parasites and their hosts. If anything, as regards the first test there is a bias *against* the Hypersea theory.

2. *Internal parasites (endoparasites) that interact with one another, either attacking or joining forces with other endoparasites within the same host, should be more likely to occur, and be more abundant, in terrestrial hosts than in marine hosts.* For example, it is impossible for termites to digest wood without their complex and highly specialized gut microbiota. Protoctists, bacteria, and fungi join forces in the termite gut to break cellulose into its constituent sugars. On the other hand, the world's largest bacteria are found inside the gut of a marine fish.

Therefore, this prediction has not been assembled post hoc—it is possible that an undiscovered marine analog to, say, the termite gut microbiota exists.

Endoparasites within endoparasites should also be more common on land, all the way down to the level of microbes, viruses, and subviral genetic particles. In other words, intraparasite interactions will be more complex on land and the nestedness of parasitism will be more pronounced—the internalized parasite food chain will be longer.

3. *The terrestrial biota should have a greater proportion of parasitoids (host-killing parasites) and also more asymptomatic parasites than does the marine biota.* Another way of saying this is that there will be greater variance in the degree of virulence of hypermarine parasites in comparison to marine parasites; that is, all other things being equal, marine parasites will on average inflict a more constant level of damage to the host.

It has been shown that the easier it is for a parasite to move from host to host, the more like a parasitoid (i.e., more damaging to the host or more virulent) the parasite will become.[17] This is true for both marine and hypermarine parasites, but there should be more opportunities for the evolutionary changes in hypermarine parasite mobility, and thus a greater range in hypermarine parasite virulence.

Parasitoid wasps would be important subjects in any test of this prediction, since many of these wasps are hyperparasites. Chalcid wasps, in particular, are a good example (figures 8.2 and 8.3). Like the bdellovibrio bacteria mentioned in the discussion of the origin of mitochondria, chalcids are parasitoids that must kill their host in order to reproduce. These wasps, economically important in integrated pest management performed by agriculturalists, parasitize the eggs of other insects.

Multiple wasp and other parasites can also infect a single host, resulting in a Hypersea community of mixed parasites.

In what might be considered an extreme extension of the concept of symbiogenesis, a whole new system is created when members of different metazoan phyla, along with monerans and protoctists, all live together within a single host, whose body acts as a living environment. Hosts of the chalcid wasps eventually die because of the association; in other cases, such as millipedes and their internal symbionts, the host thrives.

Once again, this prediction is thrown out onto a level playing field. Parasitic crustaceans called bopyrids do harm to shrimp fisheries off the Australian coast because they damage the shrimp. Marine biologists have proposed that this damage might be contained by using a hyperparasitic crustacean called a cabiropsid. Cabiropsids attack the bopyrids, thus protecting the shrimp. The marine realm has a number of hyperparasitic cases comparable to those in Hypersea.[18] But for our theory to hold up, a full tally of hyperparasitism on land and at sea should give a significant edge to the land.

4. *Photoautotrophic land animals (none of which are known today or in the fossil record) that contain either eukaryote symbionts or chloroplasts in their tissues will be found to have lived at some time during the past 500 million years—they may even still be alive.* We frankly admit that the absence of photosynthetic land animals presents a stumbling block for the Hypersea hypothesis. This is because land animal tissues would seem to present an inviting hypermarine environment for photosynthesizing microbes. Photosynthetic symbionts in animals have probably been common in the sea at least since the Garden of Ediacara. Perhaps there is something about the body coverings, the water losses incurred from sitting around in the sun, or the mobile and often shade-seeking lifestyles of land animals that preclude a photosynthetic life strategy.

Overall, we believe the four predictions we offer in this chapter, plus the three we stated earlier, indicate that the Hypersea concept has the potential to become a fruitful scientific viewpoint that generates testable hypotheses with theoretical implications. This utility is in addition to Hypersea's possible contribution as a helpful way to think and talk about the land biota as a whole—as a biogeophysical force of profound planetary impact.

NINE

Hypermarine Oceanography

When the fall of dew lifted, there, over the surface of the ground, lay a fine and flaky substance, as fine as frost on the ground. When the Israelites saw it, they said to one another, "What is it?"—for they did not know what it was. And Moses said to them, "That is the bread which the Lord has given you to eat." —*Exodus 16: 14–15*

An old idea for explaining the origin of land plants, articulated by Edward W. Berry in 1945, envisages "a laying bare of the shallow sea bottom and the direct survival and modification of some of its denizens into Land plants."[1] This curious and incorrect idea has elements in common with Lysenkoism and Lamarckism, particularly as regards the implied inheritance of acquired characteristics. Indeed, it incorporates the suggestion that heritable traits can be acquired under stress. Similar expectations fueled the Soviet scientist T. D. Lysenko's belief that his wheat grains scattered over Siberian wasteland would "adapt and grow." Needless to say, they didn't. And, unfortunately, human lives depended on a harvest from that experiment.

The idea of a polyphyletic origin of land plants, emerging anew from stranded algae during a major regression of the sea, is for us a paleontologically unattractive idea, but it has an element of validity. From a hypermarine perspective, organisms living on land represent both the marine biota and the waters in which the marine biota lives.

Living organisms can move from independent, aquatic life

to symbiotic fusions of land-adapted organisms (e.g., plants and their mycorrhizae, and perhaps even the aboriginal plant as the merger of an alga and a fungus—if the Fungal Fusion hypothesis is correct). In doing so they have major geological and biological impact. Mycorrhizae do not have direct counterparts in the sea, although marine sponges form mycorrhiza-like nodules on the submerged roots of mangroves. (Research in the Mount Holyoke College Biology Department is underway to determine if there is a symbiotic interchange of nutrients between the tree and the sponge.)[2]

In this chapter we will discuss the counterparts of marine upwelling, marine currents, and marine sedimentation that occur in a hypermarine setting—that is, on land, and orchestrated by the land biota.

Hypermarine Upwelling and Biomass Enhancement

Biotic activity at the sea surface so depletes the surface water in phosphorus, nitrogen, silica, and other nutrients that there is actually an oceanographic horizon in the sea called the nutricline, above which the lack of nutrients inhibits growth of many types of marine organisms. The needed nutrients are two hundred meters down and in the dark where photosynthesis won't work. Thus, the surface of the sea is usually a low nutrient area. Such starvation is why oceanographers refer to open ocean surface water as an "aqueous desert."

Upwelling is a process that brings deep, nutrient-rich waters up to the sea surface. Upwelling is extremely important for marine life because, where and when it occurs, it alleviates the chronic starvation for nutrients by organisms trying to grow and reproduce in marine surface waters. But upwelling is a purely physical process; it does not happen everywhere, it does not happen year-round, and the marine biotas have not evolved ways to make it happen at their command, so to speak. But they have been able to work up at least one trick.

In an adaptation that partly relieves this nutrient lack, the marine diatom genus *Rhizosolenia* has been identified as a living courier for nutrients from below the nutricline. *Rhizosolenia* is a diatom (a photosynthetic unicellular marine protoctist) that lives together with many others of its kind to form tangled, floating mats. By means of an unknown physiological mechanism, these *Rhizosolenia* mats are able to alter their buoyancy so as to sink below the nutricline and then float back to the surface carrying an enhanced charge of nutricline nutrients, particularly nitrogen. The nutrient infusion is carried in the individual cells making up the *Rhizosolenia* colony. Thomas L. Hayward suggests that *Rhizosolenia* is acting as a "marine dumb waiter" by raising nutrients in this way.[3] This marine version of biotic upwelling is woefully ineffective, however, in comparison to the hypermarine upwelling that has evolved on land.

The phenomenon of hypermarine upwelling, introduced in chapter 1, is the most biogeochemically important aspect of Hypersea. For billions of years the land surface was on the starved side of a nutricline barrier that marine organisms and early land microbes could not cross. The nutrients were locked deep within and below early soils. With the assistance of their mycorrhizal fungi, however, vascular land plants not only were able to firmly establish their tenuous hold on land and gain access to nutrients locked in soil minerals and rocks but were also able to institute a hypermarine upwelling that brings the nutrients to light by a combination of nutrient absorption and upward transport. This biotic upwelling circumvents on land the major limitation to increased biological productivity in the sea.[4]

Land or hypermarine upwelling explains at least in part the discrepancy between oceanic biomass and the much greater amount of hypermarine biomass. Simply put, if more nutrients are available and sunlight is adequate for photosynthesis, the environment will support more biomass. There may also be an indirect link between hypermarine upwelling and high land diversity—greater nutrient supplies may help explain the inordinately high diversity of the land ecosystem, as compared with the sea. (As mentioned before, the Hypersea strategy of internalizing food chains surely also contributes to enhanced diversity.)

Hypermarine upwelling is of course limited in land environments such as deserts, where water scarcity limits the capabilities of hydraulically run vascular systems for directing nutrient flow. The fact remains, however, that Hypersea entrains a large amount of ground water and uses it to drive transpiration by plants, even in surprisingly dry regions of the land. The only areas on land where Hypersea is absent are areas where liquid water is very rare or nonexistent, such as glaciers.

This is not to say that Hypersea has unlimited access to needed nutrients. Carnivorous plants have evolved as many as six different times. Digestion of insect bodies provides these plants with scarce nutrients, such as nitrogen.[5]

Currents on Land

Upwelling, in the sea or on land, is the vertical movement of nutrients. What is the hypermarine equivalent of an ocean current, which accomplishes horizontal movement of fluid?

The migrations of birds and locust swarms transport Hypersea in enormous quantities over vast distances. A locust swarm has been likened to a flying mountain—but truly it is more like a flying sea. Locusts have two phases, the *solitaria* phase and the *gregaria* phase. During their *solitaria* phase, locusts work minimal damage on plants. But if the resident plants become stressed by environmental challenges like drought, the locust population explodes into its *gregaria* phase. When the already weakened plants are completely stripped of edible parts, the locusts migrate, devastating the plant communities wherever they touch down. Later, the locusts settle down and resume the stealthy *solitaria* phase.

This type of biotic transport can be considered a horizontal, Vernadskian version of *Rhizosolenia*'s vertical (dumbwaiter) nutrient transport in the sea. It is a hypermarine current. And there are many others examples—not all of which derive from organisms that are themselves motile.

Recall the giant fungus *Armillaria bulbosa*, mentioned in chapter 4. Its vast underground network of hyphae makes it one of the largest and probably oldest living things. A single organism of such a tremendous size clearly has local geological and geochemical impact on its surrounding environment. But such an organism can also generate a kind of current within the ecosystem, transferring hypermarine components horizontally between organisms of differing kingdoms over fairly large distances. The scope and scale of the transport service is in some ways more remarkable than the tremendous size of the giant honey mushroom. Unlike the situation with the diatom *Rhizosolenia*, the flow and nutrient transport is not simply *by* living tissue, but *through* living tissue.

Indian pipe, or *Monotropa*, is a colorless vascular plant lacking in chlorophyll. Therefore, it is commonly believed to be either a pure saprotroph or a parasite of tree roots. Radioactive tracers, however, show that in truth a fungal mycelium establishes a mycorrhizal association with both the *Monotropa* root system and the roots of neighboring trees.[6] The fungus is a symbiotic intermediary link between two unrelated plants, and the interkingdom nutrient transfer is quite literally a Hypersea current, a kind of Gulf Stream of nutrients from primary producer to the consumer downstream. Mycorrhizal connections such as these are common even between two or more photosynthetic organisms, often unrelated. In other words, trees in a forest are linked by their mycorrhizae, and nutrients can flow (perhaps following nutrient concentration gradients) from one part of the forest to another as needed. The plants act as nodes in a network of fungal hyphae.

Plants that propagate by runners, such as strawberries and spider plants, can of course create analogous Hypersea currents above ground, flowing though their runners; even unrelated plants can do the same when they fuse branches or roots. For

example, in the summer of 1993, in Easthampton, Massachusetts, gardener Marjorie Root (yes, Root) reported that a *Euo-
nymous* shrub (Order Celastraceae) had grafted on its own
accord, branch-to-branch, with her Rose of Sharon bush (*Hi-
biscus syriacus*).[7] Such spontaneous grafting also occurs in natural forests and is well documented in the tropics.

Below ground, vegetative propagation also occurs. Golden-
rod (*Solidago altissima*), unfairly blamed as the plant responsible for hay fever—the real culprit is ragweed (*Ambrosia*)—uses
runners from a central mother plant to form a cluster (called a
genet) of genetically identical plantlets. It has recently been
shown that the mother plant shares nutrients with its clone
daughter plantlets. The lovely quaking aspen (*Populus tremu-
loides*) is another example of below-ground connections. A
single clone of aspen trees can cover several acres in the patchy,
fire-worked slopes of the Rocky Mountains. As a connected
whole, the genet, in theory, can take advantage of sunlight in
one spot and nutrients in another.[8]

The Strange Case of *Septobasidium*

We have saved the next example for so late in this book for two
reasons. First, the strange case of *Septobasidium* is possibly
our favorite example of a striking and unexpected biological
interaction—an interaction that is likely to seem quite anoma-
lous unless one is aware of Hypersea. Second, this example
nicely illustrates the capabilities, both realized and potential, of
Hypersea organisms. We predict that interactions of the sort
described in this section will come to be seen as more common
in the living biota and in ancient land life than anyone could
have surmised. Also, we feel that the following example is
an indication of the direction Hypersea may evolve in the
geological future.

Septobasidium (figure 9.1) is a gelatinous fungus found on

tree bark.[9] It acts as a link in a Hypersea current that flows
through a scale insect as well as the tree. Scale insects, members
of the family Coccidae, suck sap from trees by means of a
strawlike mouthpart called a suctorial tube. Most scale insects
cover themselves with a protective scale or shell of wax, and
heavy infestations of these insects can occur on leaves, twigs,
and fruit.

The biotic relationship between *Septobasidium* and the scale

FIGURE 9.1

Septobasidium's insect "house" shown in cutaway view. The scale insect
has given birth to one of its free-living young, which is crawling out
through an opening or door in the house. The scale insect's suctorial tube
can be seen reaching into the tissues of the plant where it taps into
phloem. Length of mother scale insect is approximately 1 mm.

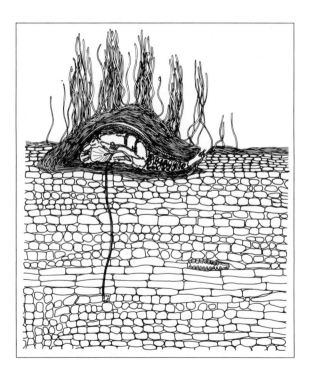

insect has been referred to as an "animal-lichen." The association is often characterized as parasitism of the scale insect by
the fungus, but it appears that the insect also gains benefit from
the relationship. The association is a mutualistic symbiosis.

Septobasidium forms a dense, flat mycelial growth on a tree
branch. The mycelial mat possesses chambers, channels, and
tunnels which harbor a colony of scale insects. Some of the
adult insects are penetrated by *Septobasidium* hyphae, which
branch into complex haustoria inside the scale insect. The
"lichenized" insects are paralyzed but not killed, and they are
able to give birth to numerous progeny. In effect, the insects
are turned into living pumps; they continue to feed on the sap
of the tree branch even as nutrients and water are passed
through their own bodies to the invading fungus.

Why do we think the insects are not just used by the fungus?
What leads us to think that the association is mutualistic?

Consider: the haustoria-invaded insects have been found to
live longer than their still-mobile relatives. In fact, the invaded
insects become pillars of their society, so to speak, since they
nourish the fungal home on which the welfare of the insect
colony depends. Young insects apparently play a role in propa-
gating the animal-lichen since they carry fungal basidiospores
with them when they set out to found new colonies.

The strange case of *Septobasidium*, then, demonstrates at least
a three-kingdom current of Hypersea nutrients and fluids;
from mycorrhizal fungus to tree root, from tree trunk to scale
insect, and from insect to gelatinous fungus. This might seem
to be an isolated and atypical example, but at least twenty
species representing nine genera of scale insects are lichenized
by *Septobasidium,* and the animal-lichen is not even restricted
to exotic climates or faraway places; it has been recorded all
over the eastern United States. Furthermore, chalcid wasps
parasitize the scale insect *Aspidiotus* as it grows in intimate
association with *Septobasidium burtii.* The wasp preferentially
lays eggs on the immobilized, lichenized insect host. Wasp

larvae and fungal tendrils coexist for a time within the body of *Aspidiotus*.

In this more complex and even stranger case of *Septobasidium*, body fluid passes from mycorrhiza, into the tree roots and up the trunk, where it is sucked up by the scale insect and then absorbed both by the fungus and the parasitic wasp larva. All this does not even consider any of the viruses, bacteria, protoctists, and even nematodes which are certainly associated with the body tissues of the interlinked fungi, plant, and animals.

Body Fluids as Community Property

The individuality of one's precious bodily fluids would seem to be a rather basic attribute of life. Indeed, the western medical tradition has viewed intrusion of this sacred space as a great evil, and this sentiment (right behind the desire for health and the profit motive) has motivated the search for vaccines or other biochemical "magic bullets" capable of wiping out offending "germs." The case of *Septobasidium* challenges the notion of normal, healthy body fluid as a sterile medium.

In Hypersea, intimate biotic interconnections are to be expected; they constitute an essential part of the land biota. In the marine realm, living things exist in a physical environment populated by other living things; in the hypermarine realm, we see living things existing in an environment that is living. Sometimes the living environment is the body of another organism, other times the living environment is manifest as an intimate chain of connected organisms.

The flow of the hypermarine current directed by *Septobasidium* is bidirectional; metabolites (metabolic byproduct chemicals) derived from the *Septobasidium* fungus are apparently injected via the insect into the plant.[10] This case cries out for some meticulous experimental studies. What influence, if any, do the fungal metabolites have on the insect? The plant? The scale insects continue to reproduce even in their lichenized state. Do the juveniles aid propagation of the fungus? Do the mycorrhizae below ground "feel" the influence of the *Septobasidium* injections into the plant, and if so are the growth, reproduction, or some other characteristic of the mycorrhizae influenced in any way?

Lest you be thinking that the *Septobasidium* example is too exceptional to permit its use as evidence for general application of the Hypersea concept to the entire land ecosystem, consider the next case.

Heterodera schachtii, the sugar beet nematode, is a dreaded pest of an important root crop. This nematode provides a possible subterranean analog to the above-ground case of *Septobasidium*. A nematode scientist has succeeded for several years running to get sugar beet nematodes to infect the roots of a tomato plant.[11] The nematode cysts on the roots of the tomato are hyperparasitized by the mycorrhizal fungus *Glomus*, the same genus we first encountered as fossils in the Rhynie Chert.

Good fossil evidence indicates that interactions between soil fungi and animals have occurred for millions of years. In an amazing and unexpected example of superb preservation, tiny fossil nematodes of the species *Oligaphelenchoides atrebora* have been found transfixed by fossil fungi. In these specimens from Cenozoic Mexican amber, the fungal spores and hyphae are growing *inside* the nematodes. The fungi appear to belong to a group of nematophagous or nematode-eating fungi. One can imagine a nematode invaded by a fungus and that was still able to move around and obtain food, or alternatively a nematode that settled down to a sedentary existence with its internal mycobiont. In any case, there seems to be poetic justice in this ancient nematode-fungus association, for *O. atrebora* belongs to a group of fungus-eating nematodes, and uninfected nematodes of the same species apparently fed upon the nemesis fungus.[12]

We see these cases as actual or potential examples of fungus-animal-plant linkages of the type encountered in *Septobasidium*, only they occur below ground. The concept of Hypersea suggests a more generic research project on this theme: Have interactions of this sort been overlooked in soils? We know that soil fungi are capable of capturing and devouring soil nematodes. Why should not the fungi "mycorrhize" them as well? For that matter, are there other soil organisms, especially sedentary ones such as white root aphids, that are also liable to be mycorrhized?

Such considerations taken to their "hyperthetical" extreme lead to a conception of a woody vascular plant, healthy and strong, but with lichenized animals and parasitic epiphytes in its branches and, for symmetry's sake, mycorrhized animals and colorless, parasitic plants attached to its roots. Hyperparasitic fungi cling to and penetrate both the fungi above ground and the mycorrhizal fungi below ground (recall the fungal hyperparasites mentioned early in this book that infect both modern mycorrhizae and fungi of the Rhynie chert). Metabolites from any one component flow freely between the living links in the system, and although at first glance many members of the consortium appear to be beset with a heavy burden of hangers-on, in fact most of the parasites and especially the mycorrhizae make beneficial contributions to the core plant, contributions that tend to encourage perpetuation of the system as a whole.

This hypothetical example of what we fantasize as the ultimate hypermarine community has not yet been described in nature, but we attribute this absence not to a deficiency in nature but to a deficiency in the imaginations of researchers. Quite simply, the ultimate hypermarine communities (or at least certain aspects of them) have not been found because nobody has looked for them.

The immune systems and other defenses of the majority of land organisms struggle tenaciously, in classic darwinian fashion, against the intrusions of parasites. But these defenses are so frequently breached that a situation akin to our fantasy example is certainly not out of the question. Artificial genetic manipulation could even allow the intentional engineering of a similar degree of interconnections, should agriculturalists find them useful. "Hypermarine aqueduct" might be a good name for such a creation.

The scientific analysis of natural hypermarine connections is just getting under way. Hypermarine linkages have been identified between the roots of the willow *Salix repens* and its mycorrhizal fungus (a basidiomycete), along with the chlorophyll-lacking parasitic plant *Monotropa* and the flowering plant *Pyrola* and possibly the liverwort *Aneura*. All are linked together by the fungus, forming a hypermarine daisy-chain. The fungus appears to be the glue that holds the daisy-chain together.[13]

Hypermarine Sediments and Evaporites

Hypermarine sediments are extremely rich in carbon—far more so than are their marine counterparts. One vast depositional basin for hypermarine sediments occurs in the trunks of trees. Each tree ring can be thought of as a hypermarine sedimentary lamina that forms over the course of a year, during which time the tree in essence refreshes itself with an entirely new set of cells. Inner wood can be thought of as a lifelike substrate or perhaps even "living sediment." In a humid climate, like that of our own Massachusetts, mushrooms and toadstools spring up so quickly from the woody roots of felled trees that some of them must certainly have come from mycorrhizae of the living tree. Once the living source of food is gone, the fungi apparently divert their resources into reproduction. They may also opportunistically act as decomposers, or they may leave that to other fungi.

Given the proper physical conditions, hypermarine deposition of sediments in the woody parts of vascular plants may lithify (fossilize) into coal formations. Coal forms whenever plant material accumulates in environments protected from oxygen and bacterial decay. Bacteria have a hard time decaying wood even when it is exposed to water and oxygen, and if the wood or other plant matter is buried under enough sediment, it has an excellent chance of becoming coal. Hypermarine production of this refractory (i.e., resistant to decomposition) material became geologically important in the mid Paleozoic and has been so ever since, particularly during the Carboniferous.

Jennifer M. Robinson has argued that more coal was formed during the Carboniferous than at any other time before or since because there was a lag between the abundant production of plant matter and the appearance of organisms (principally fungi) capable of rapidly decomposing such accumulations. This lag led to a major bottleneck in the recycling of fixed terrestrial carbon, resulting in the massive accumulation of fossil organic material as the world's great coal deposits. The accumulation of lignin and other refractory organic matter thus was something of an "oxygen crisis" of the land and also had a significant geochemical impact. Robinson's hypothesis, however, requires careful scrutiny, because coal deposits are known to be shot through with fossilized fungal remains.[14] Coal is the long-term storage form of hypermarine sediment, whereas wood and other tissue of living plants is the short-term storage. Both types are eventually reworked back into the global carbon cycle.

Hypersea contains not only the terrestrial equivalent of marine sediments but the terrestrial equivalent of marine evaporites. Marine evaporites are salt and mineral deposits formed in seas with high rates of evaporation and restricted circulation, such as the Mediterranean. Hypermarine evaporites are born of the same fundamental physical conditions. For example, anthropogenic hypermarine evaporites are made right here in New England.

Perhaps the nicest thing about living in New England, especially during years when the spring nights are cold and days are warm, is the abundance of maple syrup. A gallon of maple syrup requires evaporation of ninety-nine gallons of water from one hundred gallons of tapped tree sap. Maple sugar candy can be made by dehydrating the syrup even further. Maple syrup and candy are truly hypermarine evaporites, albeit man-made.

A natural hypermarine example, combining elements of evaporites, currents, and sedimentation in Hypersea, is described in an ancient source, the Bible. The story of the miraculous food eaten by Israelites wandering in the desert for forty years is probably based on a natural phenomenon. In Wadi Sheikh in western Sinai, tamarisk trees grow in profusion. In June the tamarisk trees (*Tamarix mannifera*) release what is understood by the local residents to be manna. Manna falls to the ground in small drops, where it dries. It is actually gathered up in accordance with the description given in Scripture. The harvest usually lasts six weeks but only about five hundred pounds total are gathered in a season. It must be sought before the sun comes up, or else it reliquifies.[15]

Manna is found only when there is abundant rain. The Bedouin Arabs preserve manna in leather gourds, saving it like honey. It is brownish yellow in color and has a pleasantly sweet but spicy taste. The Arabs consider it to be the first delicacy yielded by the land.

Manna is now known to be the exudation of several species of scale insects (one is *Trabutina mannipara*) that live on tamarisk trees. Chemical analysis of manna shows it to be composed of three simple sugars plus pectin. The plant sap on which the insects feed is rich in carbohydrates but contains very little protein. Plant tissues and fluids are relatively nutrient poor, and are especially low in aromatic amino acids. In order to obtain the amino acids needed for their metabolism, the insects pump great quantities of sap and excrete the carbohydrate-rich excess. In the dry desert air, the honeydew secretions quickly dry into sticky solids. Thus manna is a naturally formed Hypersea evaporite deposit.

The ejection of the sugar-rich extract of phloem or xylem is a common phenomenon in the insect world. The puzzle of the

Mexican raintree, underneath which it "rains" even on a sunny and dry day, was solved when it was shown that the trees were covered with numerous "sharpshooter" cicadas. The cicadas periodically eject sprays of dilute honeydew.[16]

In a process recalling the making of maple syrup, residents of some parts of the Middle East are able to concentrate phloem exudate without the help of insects. They do this by making slanted cuts in the bark of ash trees, and then collecting the hardened sap that forms in the heat and aridity of the desert atmosphere. The exudate forms sugary rods called cannelli.[17]

Other hypermarine evaporites are formed by plants without animal intervention. Through a process called guttation, plants sometimes expel excess water from their leaves, forming dew. The pores through which this water passes often excrete salts as well as water. The leaves of plants of the saxifrage family (for example the genus *Saxifraga*) frequently develop a whitish crust composed of calcium salts. This saxifrage leaf crust, formed as dew evaporates, qualifies as a very thin hypermarine limestone deposit.[18]

Hypersea as a Biogeological System

The main implications of the Hypersea view of the land biota are as follows. First, because of its inherent compartmentalization, the Hypersea biota is likely to be much more transportable than the marine biota. We mean this in the sense of transport as "space biospheres" or terrarium-like colonies on other planets. Better to take a bonsai maple tree and its mycorrhizae to the moon or Mars than to try to take a scaled-down coral reef.

Discrete oases of Hypersea can survive quite well in isolation. A century ago, Russian Pyotr A. Kropotkin was inspired by a Siberian journey to see the natural world in a new way—which he recounted in his now-classic book, *Mutual Aid*.[19] Specifically, Kropotkin was impressed with the interrelationships linking the organisms in mutualistic associations within isolated patches of taiga forest. He felt that in a physically harsh environment organisms were better characterized by their tendency to join together in their struggles to survive the inclement conditions than by the then-common portrayal of darwinism as a bitter Tennysonian struggle among organisms, "red in tooth and claw."

Mutual aid theory thus regards cooperation as more important than competition in nature, the general idea being that unless organisms assist one another, they will be unable to cope with the difficulties of their physical surroundings. The theory has idyllic connotations that accorded well with Kropotkin's anarchic political views.

Without embracing the idyllic cast of the mutual aid theory, we have found Kropotkin's worldview useful in helping us understand the implications for Hypersea of what Russian scientists after Kropotkin came to call "geobiology" and "symbiogenesis." The creatures of Hypersea have had over 400 million years of experience constructing, packaging, protecting, and dispersing an internal aqueous environment, and in the process have created a new and significant geological entity.

Why did this process begin 400 million years ago? We suspect it was related to the Cambrian "explosion" and to the radiation of aggressive, brain-directed metazoa, as discussed in our 1990 book, *The Emergence of Animals: The Cambrian Breakthrough*. The Cambrian explosion of abundant animal life is the real beginning of the noösphere or "thinking layer" of the earth, which has joined the biosphere, the atmosphere, the hydrosphere, and the lithosphere as recognizable features enveloping our planet. (LeRoy invented the term noösphere; Vernadsky[20] gave it scientific currency.)

In retrospect, the most important aspect of the Cambrian event for the land biota is the appearance of large, brainy predators in the sea. Green algae lived for billions of years during the Proterozoic, but it was not until shortly after the Cambrian explosion that they were induced, by animals wanting to eat them, to breach the air barrier. Forms such as *Palaeonitella* of the Rhynie Chert were likewise pressed by grazers into marginal aquatic environments with fluctuating or even nonmarine salinity levels and periodic conditions of desiccation. Such conditions would surely have been injurious to many grazing metazoans of the early Paleozoic. Fluctuations in marginal marine aquatic salinities may have fortuitously prepared the ancestors of plants to survive the desiccation-induced osmotic stresses their descendants would later encounter in the atmosphere.

Green algae might not ever have colonized these marginal marine environments unless forced, via natural selection, to do so by Cambrian and later types of efficiently grazing, brain-driven animals. We do not mean to imply here that the green algal ancestors of plants intentionally moved away from threatening marine grazers and environments and onto land, only that the activities of grazers led to a change in the genes of these green algae, with the genes coding for survival in air being preferentially favored (and in a sense, concentrated) by the feeding activities of subtidal animals. And when specifically the charophyte green algae developed drought-resistant oogonia, and other green algae modified morphology or lifestyle in accordance with pressures of natural selection in freshwater environments, they and other groups were in a position to "discover" the advantages of life in the air.

The combined influences of environmental rigor, predation pressure, and major new photosynthate supplies afforded by sun in air led to the evolution of a new kingdom of land organisms—the embryophytes or plants. Assuming true fungi had already evolved in the sea, plants were the last and greatest kingdom to be added to Earth's biota, and they originated on land. (If the Fungal Fusion hypothesis is correct, the kingdom Fungi would have *had* to evolve before the kingdom Plantae, because Plantae would have originated through the fusion of an alga with a fungus.) As complex life colonized land and simultaneously colonized its own tissues, it spread Hypersea over the land surface, like a rising tide or a transgressing sea.

Hypersea is not merely a novel venue for conventional biological activity, however. It is a new and very lively biogeological system. If airborne propagules such as fern spores and the pollen of later plants have a decidedly horizontal, Vernadskian motion, then in contrast the hypermarine upwelling seen in vascular plants has a uniquely vertical aspect, as well as an importance for all the other organisms living on land. This importance has no living parallel in the sea.

By providing refugia for organisms like the parasitic pentastomes, now permanently absent from the oceans, Hypersea is a major force in maintaining the biodiversity of the global biota. More important, Hypersea provides unique habitats for the evolution of uncounted species of viroids, viruses, microbes, fungi, animals, and parasitic plants that live wholly within other terrestrial organisms. These interior inhabitants of Hypersea actually constitute the bulk of Hypersea's contribution to the global species tally.

One final example can further illustrate the ability of hypermarine interactions to maintain diversity of species on land. The American Chestnut (*Castanea dentata*) is a stately tree that can reach a height of over thirty meters. In 1904 the Chestnut Blight Fungus (*Cryphonectria parasitica*) was accidentally introduced to North America. Native to Asia, where it parasitizes but does not kill Asian chestnut tree species, the blight fungus had a devastating effect on the immunity-lacking American species. By World War II, American Chestnut trees were nearly extinct. Shoots continue to sprout from the stumps of former trees, but quickly succumb to the blight. A few— but only a few—carefully cultivated American Chestnut trees continue to thrive in areas where the blight is absent.

This might appear to be a case in which a parasite-host Hypersea interaction leads to a loss in diversity, namely, loss of *Castenaea dentata*. A new development, however, indicates that this loss will probably not take place. Instead, it appears that ultimately a gain in diversity will be realized.

The story of this new development begins with a process called anastomosis. When two or more different strains of blight fungi infect a single tree, hyphal anastomosis can occur. During hyphal anastomosis, fungal hyphae fuse and undergo cytoplasmic exchange—a thoroughly hypermarine sort of event. This phenomenon is now, apparently, under way and may eventually bring the American Chestnut back from the brink of extinction. This is because the blight fungus is hyperparasitized by viruses. One virus, called a dsRNA virus (double stranded ribonucleic acid virus) is believed to be the genome of a defective virus unable to produce a protein coat. It is nevertheless quite capable of hyperparasitizing *Cryphonectria parasitica*. This unusual virus induces a condition in the fungus called hypovirulence. That is, although the fungus continues to infect the American Chestnut tree, a fungus infected with the virus no longer causes the unfortunate symptoms that damage the tree. If an American Chestnut infected with virulent blight is later infected by blight carrying the dsRNA virus, the two blight fungi undergo hyphal anastomosis (as long as the two fungal strains are compatible). The dsRNA will be passed to the virulent fungus, which is then rendered hypovirulent. The chronic symptoms of blight disappear from the chestnut, and the tree may ultimately restore itself to health.[21]

Here then we see a situation in which we have three species (tree, virus, and two strains of fungus) where, without the existence of the double stranded RNA virus, we might have had no species at all, owing to local extinction of the American Chestnut. All this is without regard to any other strains of fungi, viruses of the fungi, or any other pathogens or symbionts of the tree itself which are likely to be present. Instead of a diversity loss, we see a maintenance of diversity arising out of Hypersea's proclivities to treat the body fluids of one organism as a community resource.

TEN

Humans as Reservoirs of Hypersea

Despite advances in Western medicine in the past few hundred years, access to our own tissues by other organisms remains a reality. We humans are very much a part of Hypersea.

Our Inner Biota

In chapter 8 we discussed the tragic case of Roberta Hanson and the parasitic nematode *Ascaris lumbricoides* that strayed into her brain, bringing with it a contingent of normally friendly gut bacteria that wreaked havoc in the exotic habitat. Another incident came to light recently that originally had doctors at the New York University Medical Center quite puzzled. A six-year-old boy from a Jewish family in Brooklyn was rushed to the hospital after he was stricken with a seizure. At first, doctors feared a brain tumor, but a CT scan revealed the real cause. The scan found two cysts in the boy's brain. The cysts were the larval stage of *Taenia solium*, the pork tapeworm.

What on earth could cause an outbreak of pork tapeworm among orthodox Jews, whose religion strictly proscribes the eating of pork? As it turned out, the boy's family had employed sixteen housekeepers from Central America, where tapeworm infections are common. The housekeepers, unaware of being infected with tapeworm, inadvertently contaminated the family's food with tapeworm eggs. The eggs can lodge under fingernails and on skin, thereby spreading the infection. The boy had become infected without any violation of his religious precepts regarding consumption of pork.[1] Fortunately, tapeworm infection is easily cured with a single dose of the drug praziquantel.

Much less easily cured is malaria. Several types of malaria exist. The most common variety is caused by *Plasmodium falciparum*, a unicellular protoctist that lives in the tissues of 250 to 400 million people and causes more than a million deaths worldwide each year. A vigorous search is underway for a vaccine to prevent malarial infection.[2] Another approach might be to locate a hyperparasitic microbe that can immobilize falciparum malaria—this approach would constitute a form of integrated pest management aimed at our own bodies instead of our fields and orchards. Such a disease-combating strategy might seem far-fetched; in a Hypersea context, however, it is the obvious course of action. To our knowledge, such an approach has never been attempted. Such an attempt would have to proceed with great care, as examples to come will demonstrate.

Not all inhabitants of the human bays and inlets of Hypersea are pathogenic. The mite family Demodicidae contains a single genus *Demodex* with many species. One of these, mentioned in chapter 1, is *Demodex folliculorum*, the follicle mite (figure 10.1). These mites live in human hair follicles around the nose and eyes. The life cycle of *D. folliculorum* takes place entirely within these hair follicles. The mites are worm-shaped and only 300 to 400 microns in length. You may be infested with these mites at this moment and not know it, for they are usually asymptomatic. As much as 50 percent of the world's population harbors the tiny arthropods.[3]

More invasive and less benign is a trematode (a type of parasitic flatworm or fluke) that causes a disease called schistosomiasis, which has very unpleasant symptoms in humans. The culprit is the blood fluke *Schistosoma mansoni*, a trematode with a complex life cycle that uses the freshwater snail *Biomphalaria glabrata* as an intermediate host and humans as the definitive host. It has been suggested that young flukes adsorb the host's defensive chemical substances onto their surfaces to escape immunologic detection by the host. Biomedical researchers have invoked the standard hope for a vaccination or "magic bullet" that would stimulate human immune systems to wage war on this invasive, but immunologically invisible pathogen. Bolstered by the successes of vaccines in preventing polio and in eradicating smallpox, the biomedical research community includes many practitioners who believe that there must be a magic bullet vaccine or cure for every disease. Such remedies, however, have proved elusive in the case of schistosomiasis—and AIDS (which will be discussed later).

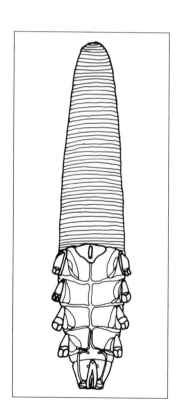

FIGURE 10.1

Demodex folliculorum, the human follicle mite (ventral view). Length of mite is approximately 300 microns. Chances are that a population of these arthropods uses your skin for habitat.

How the trematode initially evades detection by the host's immune system is still not fully understood. The mobile larvae, called cercariae, are the stage of *Schistosoma* infective to humans; they bore into a person's skin to gain entry into the blood. Once a schistosome trematode successfully establishes itself within the host, however, it gives its host resistance to infection by other schistosome larvae, which will be attacked by the host's immune system if they try to enter.[4] Such "no vacancy" messages to other trematodes obviously regulate the number of schistosome strains per host by keeping out unrelated members of the same species.

Regulation of parasite density has been studied in other trematodes. Individuals of the fluke *Zygocotyle lunata* vary in size within a single host, with size being inversely proportional to the number of individuals present.[5] This phenomenon, called the *crowding effect*, is observed in other trematode parasite populations as well as in cestode infections (cestodes are also parasitic flatworms). The crowding affect in parasites is similar to the well-known limitations to fish size in an aquarium—an artificial ecosystem. (Compare the size of a child's goldfish to those in a pool in a Japanese garden—or to the escapee carp that are the bane of sport fish enthusiasts.) Aquarists plan for one inch of fish length per gallon.

In 1934 the Danish zoologist C. J. Wesenberg-Lund reported that trematode larvae (in this case trematodes called echinostomes) will prey on other larvae and sporocysts (resting stages) within their snail host. This is reminiscent of the strategy employed by the wasp larva *Venturia* to keep competitors out of the body of its moth host, an example of hypermarine territoriality described in chapter 1 (figure 1.5). Wesenberg-Lund's observations have actually been considered as a possible approach to biological control of human-infecting schistosomes in Southeast Asia. Hyperparasites have also been proposed as possible biologic control agents for schistosomiasis and other human diseases, as well as for agricultural parasites. This control would not be a strictly Hypersea phenomenon, how-

ever, because the snail intermediates of *Schistosoma* are freshwater aquatic organisms, making *Schistosoma* sort of an "anadromous fish" of Hypersea.

The microsporan protoctist *Nosema* is perhaps the world's
most versatile hyperparasite. It is the only known hyperparasite
genus that inhabits marine, freshwater aquatic, *and* hypermarine habitats. Also, it is unusually adept at switching hosts.
Nosema algerae, normally a mosquito parasite, will become a
hyperparasite of *Schistosoma mansoni* larvae if the infective stage
of *N. algerae* is fed to the snail *Biomphalaria glabrata.* As noted
above, this snail is a favorite intermediate host of *S. mansoni.*
Hyperparasitized *S. mansoni* larvae in snails do not produce as
many mobile larvae; thus *N. algerae* is a possible control agent
for schistosomiasis.[6]

Toxoplasma and the Marburg Virus

The protoctist *Toxoplasma gondii* has some similarities with the
parasitic organisms mentioned in the previous section. Unless
you are a lifelong vegetarian, *Toxoplasma* may currently be
inhabiting your body fluids. Even then, if you are affectionate
with a cat you may have become infected. *Toxoplasma* is widespread in wild animals, domestic animals, and humans, and it
is transmitted in the food chain primarily through scavenging
and carnivory. The sexual phase of its life cycle, however,
occurs only in cats. Like *Demodex, Toxoplasma gondii* is typically asymptomatic, which is fortunate indeed as 15 to 85
percent of adult humans are chronically infected with it. Preg-
nant women are told to avoid cat litter boxes because onset of
Toxoplasma infection during pregnancy can cause congenital
disease of the newborn.

Recent research shows that, in spite of its ability to reproduce sexually, the disease-causing (virulent) strain of *Toxoplasma gondii* comprises a single, widespread clonal lineage.
This lineage has remained genetically homogenous as it has

invaded the tissues of many unrelated species of land vertebrates.[7]

For people infected with the AIDS virus (either HIV-1,
common in America and Europe, or HIV-2, endemic to Africa), *Toxoplasma* can be extremely dangerous. Thirty to fifty
percent of AIDS patients chronically infected with the parasite
develop encephalitis (inflammation of the brain) caused directly
by *Toxoplasma*. *Toxoplasma* is only one of a number of organisms that characterize the AIDS Related Complex (ARC) diseases caused by AIDS itself. Some doctors have expressed
surprise that only a fraction of AIDS patients infected with
Toxoplasma have gone on to develop toxoplasmic encephalitis,
considering their compromised immunological state. This
means that even virulent strains of the parasite cause disease in
a relatively small number of people; in other words, a balance
is usually struck between *Toxoplasma* and the immune system
of the human host. In an evolutionary sense, we feel that this
must mean that species that harbor asymptomatic *Toxoplasma*
have probably had many millennia of exposure to the organism,
allowing populations of host species time to adjust to its
presence.

Serious epidemics occur when the process of host-parasite
adjustment is short-circuited, as apparently has occurred with
the AIDS epidemic (see next section) and perhaps with the
lesser known Marburg epidemic. There is even evidence that
increasing the number of opportunities for a parasite to infect
new hosts can cause a species or strain to become more virulent
and thus more dangerous. In other words, the easier it is for a
parasite to hop from one reservoir of Hypersea to the next, the
less it is to the evolutionary advantage of the parasite to be
"nice" to the host and, as mentioned in chapter 8, the more
serious are the disease symptoms.

Marburg is the ancient German city where theologians debated religious doctrine four centuries ago. In August 1967,
twenty Marburg citizens fell seriously ill. Their symptoms
included internal hemorrhaging, vomiting, and high fever.

Five of the victims died after lapsing into comas. Their mysterious illness was soon labeled Marburg disease.

Attempts to isolate a viral or bacterial pathogen by conventional laboratory techniques failed to identify any known agents. Soon after the beginning of the epidemic, however, it was realized that most of the victims were employees of vaccine and antiserum laboratories. In the 1960s and 1970s, thousands of vervets (green monkeys) had been shipped from Uganda to Germany for the purpose of vaccine and antiserum preparation. Tissues from these monkeys were used to culture a number of common viruses. The Marburg pathogen turned out to be a virus native to these monkey tissues. In the confined laboratory setting, the virus easily infected laboratory workers who handled the vervet tissue.

Thus, the laboratory monkeys fell under suspicion as the source of the virus. But sporadic human cases of Marburg disease, without any evidence of wild animal contact, were subsequently reported in South Africa in 1975 and in Kenya in 1980. Since some instances of the disease were obviously not related to laboratory exposure, the actual source of the virus remained a mystery. One scientist remarked in dismay that "the green monkey was a red herring."

The Marburg virus is now known to be a type of virus called a filovirus.[8] It was probably derived from the laboratory monkeys. This type of filovirus is normally harmless in monkeys. Humans have little immunological protection against the virus, and the virus was apparently able to spread quickly.

Artificial breaching of the barriers between Hypersea reservoirs can thus have serious consequences. Like the lampreys introduced into the Great Lakes that attacked the native lake trout, new organisms introduced to a Hypersea community can seriously disrupt its balance.

AIDS, Viroids, and the Mixing of Alien Hypersea Fluids

Green monkeys are delightful animals to observe, and we spent many hours watching them at the zoo when our eldest daughter was young. But vervets are suspected to be the original source of HIV (Human Immunodeficiency Virus), which most researchers believe to be the cause of AIDS.[9] HIV comes in two known varieties (HIV-1 and HIV-2). Another possible virus (as yet unidentified, but dubbed "isn't") is hypothesized to cause AIDS symptoms in people carrying *neither* HIV-1 *nor* HIV-2.

African green monkeys captured from the wild have a 60 percent incidence of Simian Immunodeficiency Virus (SIV). SIV is similar to HIV, but SIV appears to be asymptomatic in green monkeys. However, when it is injected into Asian macaque monkeys it results in an AIDS-like disease.

Several hypotheses have been proposed to explain how the precursor of the HIV virus might have made the leap from monkey tissue to human tissue. These hypotheses offer suggestions ranging from monkey hunters who carelessly cut themselves to western medical scientists experimenting on other humans in Africa by injecting them with monkey blood. One provocative idea involves a practice of the Idjwi tribe. The tribe resides on an island between Rwanda and Zaire. According to verified reports, men and women of the tribe stimulate themselves to intense sexual activity by injecting monkey blood into their pubic areas, thighs, and backs.[10]

Whatever the original means of transmission, it seems that the AIDS pathogen as well as the Marburg virus may be a result of interspecies transfer of Hypersea fluid. Such a mode of transfer is clearly a possibility for *Toxoplasma* too, although its usual mode of transmission is by the eating of infected meat. *Toxoplasma* has relatively mild effects on most of its hosts, probably because most carnivorous mammal lineages have had periodic exposure to *Toxoplasma* for millions of years. In the case of HIV and Marburg, it is likely that *Homo sapiens* had never been exposed to these viruses before this century. So the viruses, like rabbits and prickly pear cactus (*Opuntia*) introduced to the Australian outback, ran amok in a new and defenseless hypermarine oasis: human bodies.

A class of disease agents with intriguing similarities to AIDS and Marburg disease can be documented for land plants. This time the pathogens are neither protoctists nor viruses, but tiny viruslike objects called viroids. Viroids differ from viruses in that they are much smaller and in that they have vanishingly small amounts of genetic material. Viroids are the smallest known agents of infection and in some cases are apparently protected by a partial protein coat that makes them invulnerable to enzymes that normally dissolve DNA or RNA. One kind of viroid causes spindle tuber disease in potatoes; another is responsible for the disease cadang-cadang ("dying-dying") that decimated the coconut crop on Bicol Peninsula in the Philippines in this century. Most plant viroid diseases seem to be regulatory afflictions confounding a plant's ability to grow; a characteristic symptom is severe stunting. It has been suggested that a viroid may originate as a normal regulatory gene in one plant.[11] This gene somehow gets into the vascular fluids of another plant species. The DNA fragment would of course also need to be able to replicate within the host plant and it would have to be capable of transmission from host to host. We feel that viroids will eventually be shown to be asymptomatic parasites (perhaps originally derived from regulatory genes) in an original host but which then ran wild when they reached the fluids of new species.

Many viruses are thought to be quite closely related, genetically speaking, to their hosts. Viral transfer between hosts, however, is a reality of Hypersea. Many human influenza viruses are Asian in origin because of the close association in Asia of domestic ducks and pigs. Avian viruses are passed to pigs and thence to humans, and a flu epidemic quickly spreads to other continents. Sap-feeding insects are important vectors of plant parasites. It has been suggested that, because of the one-to-one correspondence between plant viruses and the insect vectors that carry them, the viruses were originally *insect* viruses. Indeed, experimental study of the virus that infects both the aster-yellow plant and its insect parasite (*Macrosteles fascifrons*) has shown that the virus multiplies asymptomatically within the body of the insect.[12] Far from being a case of simple contamination, the aster-yellow virus is a stellar example of a natural, hypermarine virus transfer.

Overall, Hypersea fluids from normally alien sources that mix within the vessel of a living organism can have dangerous and unintended consequences in the same way that introduction of exotic species can forever alter the ecological character of an isolated marine or freshwater habitat. This perspective on mystery pathogens not only flows out of our Hypersea idea; the evidence we have cited tends to confirm our view of Hypersea as an extended, living environment.

The Evolutionary Potential of Hypersea

The evolutionary potential of the Hypersea environment is tremendous. If speciation normally occurs as a result of the geographic isolation of subunits of species populations (this is the conventional explanation provided by evolutionary biologists for the origin of species), then Hypersea offers *tremendous* opportunities for evolution. There are two reasons. First, every reservoir of Hypersea fluids that exists in any individual land organism is potentially a site for speciation. Second, the immune systems of hosts—which actively resist new introductions—provide an exceptionally challenging form of selective pressure for symbionts and parasites. Natural selection is made to work overtime in these internal environments.

The possibilities for evolution in Hypersea are even more impressive when one considers the etiology of AIDS. A simple but compelling model has been presented to account for the pathogenicity of HIV-1 in terms of the interaction between populations of the AIDS virus and the host immune system. This model is based on a single primary assumption. The assumption is that HIV produces antigenic variants or "quasispecies" at a high rate, and that the most recent variants keep ahead of the host immune response specific to earlier quasispecies. The host immune system can successfully respond to the growth in quasispecies for only a finite amount of time. This time interval, according to the model, is the interval during which the AIDS disease remains in quiescence. But after a certain threshold number of HIV quasispecies have been produced, the host immune system reaches a limit beyond which it cannot keep the additionally "evolved" quasispecies in check. The patient's immune system is simply overwhelmed.[13]

This view of the role of quasispecies in pathogenicity may also explain why other agents of disease (the usually asymptomatic *Toxoplasma*; also certain bacteria, fungi, and viruses) suddenly become opportunistically virulent and cause infection when AIDS begins. A recent French study[14] shows that even the usually harmless *Demodex* mite of human hair follicles can cause serious problems for AIDS patients because of their immune system breakdown. An AIDS patient's immune system is rendered ineffective and the tragic sequence begins. The checks and balances of an ordered hypermarine ecosystem are removed and death of the host ensues. In too many cases, the virus will have had an opportunity to spread to other human hosts by sexual intercourse, needle sharing, or blood transfusions. The viral lag, the time during which quasispecies accumulate, is like a time bomb, a useful strategy from the virus's point of view, for exploding through the human geography of Hypersea.

In an attempt to create a vaccine that provides protection against AIDS, Ronald C. Desrosiers and his colleagues at Harvard Medical School found that SIV-susceptible monkeys injected with a genetically modified form of the simian AIDS virus (SIV) were shielded from infection by deadly SIV. The modified strain of simian virus is altered by removing a gene sequence called "nef," a sequence whose function as of this writing is not well understood but seems to be necessary for virulence. Both SIV and HIV viruses carry the nef gene.

The modified SIV virus is active ("live") when it is injected into a healthy monkey. Perhaps it confers immunity in the same way that the schistosomiasis trematode confers host resistance to infection by other schistosome flukes—that is, by posting the biochemical equivalent of a no vacancy sign. The modified nefless SIV replicates poorly in the monkeys' tissues, but it remains alive and apparently renders the animals immune to SIV infection.[15] Derosier's experiment may be a successful instance of deliberate hypermarine engineering. Needless to say, we hope this approach succeeds—it holds the promise of becoming a magic bullet that keeps on shooting.

Certainly the types of genetic shenanigans involved in the etiology of viroids and HIV quasispecies are not restricted to the inhabitants of Hypersea eukaryotes. Similar viral transformations and invasions of new hosts are known in bacteria (even thermophilic bacteria) and in numerous marine organisms.[16] But in Hypersea, the combination of an occasionally extreme environmental patchiness plus the selective pressures provided by the highly developed immune systems of hosts carries the evolutionary process to an entirely new level. Hypersea, Earth's second newest geophysical entity (human industrial sprawl is the latest), has thus also become the site of accelerated and extremely localized evolution.

The Research Potential of Hypersea Theory

In the fields of paleontology, geology, evolutionary biology, ecological science, and even medicine, Hypersea theory opens up expansive new areas for exploration and research.

First, with respect to paleontology: The oldest direct fossil evidence of ancient Hypersea is Devonian fungi and protoctists preserved within fossil plants from the Rhynie deposits in

Scotland (figures 1.2, 1.3, 6.55, 6.56). The heightened importance of such finds becomes apparent when considered in a
Hypersea context. Further instances of such fossil associations
should be a very attractive new target for paleontologists interested in the history of Hypersea.[17] We urge paleontologists to
keep an eye out for unusual examples of hypermarine interactions in the fossil records of land organisms yet to be discovered
or already available in museum drawers and cabinets.

With respect to evolutionary biology, a major hypothesis in
need of testing (via the fossil record, via studies of living
plants, and via comparisons of gene sequences) is the idea that
associations of green algae and fungi, followed by associations
of plants and mycorrhizae, led to the origin and diversification,
respectively, of the land biota. Further, researchers should
look for "lost" hypermarine associations, such as the multiple
mycobiont associations hypothesized in chapter 6 for the origin
of spongiophytes and *Prototaxites*.

The discipline of genetics is the province for testing one
final prediction that flows out of Hypersea. After correcting for
any biases due to differences in the numbers of marine versus
the numbers of hypermarine organisms, *we predict that horizontal gene transfer or lateral gene transfer will prove to be more
common, and perhaps much more common, on land than in the sea.*
The example of the hypovirulence agent (dsRNA) in Chestnut
Blight Fungus gives a glimpse of why and how this might be
the case. If defective viruses can pass between different organisms, why not more-or-less naked genes themselves? Many
types of genetic material are capable of integrating into a host's
genetic code, and the conduits provided by Hypersea lead us
to predict that such occurrences should be very widespread
on land.

Those wishing to preserve global biodiversity should devote
special attention to protecting unique Hypersea organisms. The
southeast Asian plant genus *Rafflesia*, for instance, forms the
world's largest flower (up to well over a meter in diameter) and
is an exemplary inhabitant of Hypersea. The plant has become

so specialized for its hypermarine habitat that it lacks leaves, stems, or roots. The only vegetative part of the plant, directly connected to the flower, are the fine filamentous hairs that penetrate and parasitize tropical vines—vines which are themselves parasitic on other vascular plants. The host vines are not badly injured by this parasitism.[18] Sadly, though its range includes Borneo, Sumatra, Java, and Peninsular Malaysia, *Rafflesia* is one of the rarest of vascular plant genera, and redoubled efforts to protect this flower ought to be undertaken in light of its importance for hypermarine studies.

Sedimentary strata have been likened to the pages of a book of earth history. With Hypersea, geologists can unlock the genuine "meaning" of coal and can add the land biota to their list of geological surface phenomena that affect landforms and global chemistry. We are not asking geologists to blur the distinction between a nonliving rock and a living land organism. Geologists should, however, appreciate the real reason for the formation of coal and the powerful role played by land organisms in the hydrologic cycle and rock weathering—and thereby their influence on regional and even global climate and the productivity of oceans. We believe that the most effective way for geologists to make these shifts in geological perspective is from the platform of Hypersea theory. For example, tree ring dating studies should be viewed as hypermarine stratigraphy and sedimentology. We can even imagine a research program designed to correlate hypermarine laminae with fine laminations in marine or lake mud deposits.

As discussed in the first part of this chapter, medical researchers may be able to use insights gained from Hypersea theory to find biologically assisted ways to cure or prevent diseases that now pose or may soon pose major health problems. This is because many contemporary health threats have a hypermarine aspect—that is, they owe in part to the fact that body fluid is, to a certain extent, a shared resource. And because Hypersea calls for special attention to the under-inventoried segments of nested food chains, the hyperparasites, a Hypersea perspective may ultimately lead to significant medical break-throughs.

Evolutionary biologists should study how hypermarine evolution compares and contrasts to evolution in the sea, for evolutionary biology is the field in which the bulk of the work needed to verify or reject the predictions of Hypersea theory must take place. Ecology and paleoecology will also play key roles in evaluating these predictions. We believe that an effort to test Hypersea along the lines of the predictions we made in chapter 8 will enrich our understanding of both land and marine ecosystems. This book is only the first stage in unleashing the power of Hypersea research.

NOTES

I: Introduction to Hypersea

1. We present a technical version of Hypersea in M. A. S. McMenamin and D. L. S. McMenamin, 1993, Hypersea and the Land Ecosystem, *BioSystems* 31:145–153. Hypersea has, to our knowledge, no close precedents in previous scientific literature.

2. Heather Winkelmann, personal communication.

3. Mutualism is a type of symbiosis in which both host and symbiont benefit from the association. The fungus inside an alga inside a plant fossil is described by T. N. Taylor, W. Remy, and H. Hass, 1992, Fungi from the Lower Devonian Rhynie Chert: Chytridiomycetes, *American Journal of Botany* 79:1233–1241. The fungus in this case is technically a chytrid—a funguslike protoctist. Chytrids can be saprotrophs (organisms that feed on dead organic matter) as well as parasites. Saprotrophic chytrids have also been described from the Rhynie Chert by T. N. Taylor, W. Remy, and H. Hass, 1994, *Allomyces* in the Devonian, *Nature* 367:601.

4. For discussion of hyperparasitic fungi (mycoparasites) in the Rhynie Chert, see H. Hass, T. N. Taylor, and W. Remy, 1993, Mycoparasitism in the Lower Devonian Rhynie Chert, *American Journal of Botany* 80:89.

5. The following articles convey information on insect parasitoids: R. W. Pemberton, J. H. Lee, D. K. Reed, and R. W. Carlson, 1993, Natural enemies of the Asian gypsy-moth (Lepidoptera, Lymantridae) in South Korea, *Annals of the Entomological Society of America* 86:423–440; P. Weinstein and A. D. Austin, 1991, The host relationships of trigonalyid wasps (Hymenoptera, Trig-

onalyidae), with a review of their biology and catalog to world species, *Journal of Natural History* 25:399–433.

6. For discussion of *Venturia*, see D. C. Speirs, T. N. Sherratt, and S. F. Hubbard, 1991, Parasitoid diets: Does superparasitism pay?, *Trends in Ecology and Evolution* 6:22–25.

7. Precocious larvae in *Copidosoma* (chalcid wasps of family Encyrtidae) are discussed by M. Grbic, J. Ode, and M. R. Strand, 1992, Sibling rivalry and brood sex ratios in polyembryonic wasps, *Nature* 360:254–256.

8. For discussion of *Toxoplasma*, see L. D. Sibley and J. C. Boothroyd, 1992, Virulent strains of *Toxoplasma gondii* comprise a single clonal lineage, *Nature* 359:82–85.

9. For discussion of *Mycoplasma penetrans*, see S. C. Lo, M. M. Hayes, and J. G. Tully, 1992, *Mycoplasma penetrans* sp. nov., from the urogenital tract of patients with AIDS, *International Journal of Systematic Bacteriology* 42:357–364, and R. Y. H. Wang, J. W. K. Shih, T. Grandinetti, F. Pierce, M. M. Hayes, D. J. Wear, H. J. Alter, and S. C. Lo, 1992, High frequency of antibodies to *Mycoplasma penetrans* in HIV-infected patients, *Lancet* 340:1312–1315.

10. The association of ticks and the Lyme disease pathogen is described in A. G. Barbour and D. Fish, 1993, The biological and social phenomenon of Lyme disease, *Science* 260:1610–1616. The Lyme disease tick was previously placed in the species *Ixodes dammini* but has lately been placed in the species *Ixodes scapularis* because of mating compatibility with other members of this species.

11. For more information on the calculation of ecological efficiency, see M. Conrad, 1977, The thermodynamic meaning of ecological efficiency, *American Naturalist* 111:99–106. Michael Conrad's analysis of the regularities of ecological efficiency at the bottom of his page 105 are especially relevant to considerations of Hypersea.

12. For discussion of fig wasps and their parasite nematodes, see E. A. Herre, 1993, Population structure and the evolution of virulence in nematode parasites of fig wasps, *Science* 259:1442–1445. See also R. F. Sisson, 1970, The wasp that plays cupid to a fig, *National Geographic* 138:690–697.

13. Parasites of the *Ips* beetle are described by H. K. Kaya, 1984,

Nematode parasites of bark beetles, in W. R. Nickle, ed., *Plant and Insect Nematodes*, pp. 727–754, New York: Marcel Dekker.

2: The Interior Sea

1. Water has other properties that prove beneficial for the maintenance of life. It is highly transparent to light, and water's heat capacity (ability to store heat) is exceeded only by ammonia. No other common liquid conducts heat better than does water, nor does any common liquid have a higher surface tension. See S. I. Aksenov, 1990, *Voda i ee rol' v reguliatsii biologicheskikh protsessov*, Moscow: Akademiia nauk SSSR, "Nauka."

2. Mark's student Grace Osmer provided an estimate of the quantity of water in terrestrial bioreservoirs as 19 cubic kilometers. Dividing this by the total volume of water on Earth (1,360,000,000 cubic kilometers) indicates that terrestrial bioreservoirs hold about 1.5 billionth of the total volume of water on Earth.

3. For an estimate of forest biomass, see A. V. Lapo, 1987, *Traces of Bygone Biospheres*, Moscow: Mir Publishers.

4. "Hemo the Magnificent," Frank Capra, director, 1957, N. W. Ayer and Son.

5. See A. B. Macallum, 1926, The paleochemistry of the body fluids and tissues, *Physiological Reviews* 6:316–357. For instance, the ionic concentration of sodium (Na^+) in sea water in micromoles per liter ($mM.l^-$) is 475, whereas in the freshwater salmon *Salmo salar* it is only 328 $mM.l^-$. Macallum would have interpreted this in the following way: when bony fish first evolved, sea water had a composition approaching 328 $mM.l^-$., and since then additional salt accumulated in the ocean. Attempts (now known to be faulty) were even made to date the oceans by estimating the rate of accumulation of salt in seawater. For skeptical commentary regarding Macallum's hypothesis, see E. J. Conway, 1942, Mean geochemical data in relation to oceanic evolution, *Proceedings of the Royal Irish Academy*, B, 48:119–159; E. J. Conway, 1943, The chemical evolution of the ocean, *Proceedings of the Royal Irish Academy*, B, 48:161–212; G. E. Hutchinson, 1961, The biologist poses some problems, *American Association for the Advancement of Science Publications* 67:85–94; and C. F. A. Pantin, 1931, The origin of the composition of the body fluids in animals, *Biological Reviews* 6:459–482. A newer version

of Macallum's hypothesis has been presented by D. H. Spaarg-
aren, 1978, A comparison of the blood osmotic composition of
various marine and brackish-water animals, *Comparative Bio-
chem. Physiol.* 60:327–333.

6. See D. W. Schwartzman and T. Volk, 1989, Biotic enhancement
    of weathering and the habitability of earth, *Nature* 340:457–460.

7. Calcium carbonate and silica are geochemically linked through
    the following equation:

$$CaSiO_3 + CO_2 \rightleftarrows CaCO_3 + SiO_2$$

The reaction goes to the right when calcium silicate minerals (the
formula given is for wollastonite) weather by combining with
carbon dioxide to form calcium carbonate (limestone or its dissolved equivalent) and silica (quartz or its dissolved equivalent).
This reaction is virtually irreversible at the surface of the earth.
But when calcium carbonate and silica are buried and metamorphosed, the reaction is driven to the left and the mineral wollastonite forms, as well as carbon dioxide which is given off as a
volcanic gas. Wollastonite is a metamorphic mineral commonly
found in metamorphosed crystalline limestones.

8. For the influence of global surface temperature on evolution, see
    D. Schwartzman, M. McMenamin, and T. Volk, 1993, Did
    surface temperatures constrain microbial evolution?, *BioScience*
    43:390–393.

9. A gigaton is a billion metric tons; estimates from W. M. Post,
    Tsung-Hung Peng, W. R. Emanuel, A. W. King, H. Dale,
    and D. L. DeAngelis, 1990, The global carbon cycle, *American
    Scientist* 78:310–326.

10. On plants, carbon cycle, and weathering, see J. M. Robinson
    and R. A. Berner, 1991, Land plants and weathering, *Science*
    252:860; T. Volk, 1989, Rise of angiosperms as a factor in long-
    term climatic cooling, *Geology* 17:107–110; R. A. Berner, 1990,
    Atmospheric carbon dioxide levels over Phanerozoic time, *Science*
    249:1382–1386; R. A. Berner, 1993, Paleozoic atmospheric
    CO_2: Importance of solar radiation and plant evolution, *Science*
    261:68–70; and R. A. Berner, 1993, Atmospheric CO_2 during
    the Paleozoic: Importance of terrestrial plants and the sun, *Geo-
    logical Society of America Abstracts with Program* 25:A-82.

11. For modifications in plant anatomy due to drops in atmospheric

carbon dioxide, see K. L. Niklas, 1976, Morphological and ontogenetic reconstruction of *Parka decipiens* Fleming and *Pachytheca* Hooker from the Lower Old Red Sandstone, Scotland, *Transactions of the Royal Society of Edinburgh* 69:483–499; R. A. Spicer, 1989, Physiological characteristics of land plants in relation to environment through time, *Transactions of the Royal Society of Edinburgh, Earth Science* 80:321–329; and W. G. Chaloner and G. T. Creber, 1990, Do fossil plants give a climatic signal?, *Journal of the Geological Society of London* 147:343–350.

12. For discussion of the impact of land life on soil and rocks, see P. Krapfel, 1981, Gaian geology?, *CoEvolution Quarterly* 29:65.

13. Discussion of the link between oceanic anoxia and the spread of plants on land is hypothesized by both T. J. Algeo and J. B. Maynard, 1993, The Late Devonian increase in vascular plant 'rootedness': Source of coeval shifts in seawater chemistry, *Geological Society of America Abstracts with Program* 25:A-83, and M. A. S. McMenamin and D. L. S. McMenamin, 1993, Hypersea and the land ecosystem, *BioSystems* 31:145–153.

3: The Marine Prelude

1. For the history of the Russian symbiosis research program, see L. N. Khakhina, 1992, *Concepts of Symbiogenesis: A Historical and Critical Study of the Research of Russian Botanists*, L. Margulis and M. McMenamin, eds., New Haven: Yale University Press.

2. *Eosphaera tyleri* has been compared to eukaryotic organisms of both the green algal order Volvocales and the red algal order Porphyridiales.

3. The five separate symbiogenetic algal groups are chlorophytes, chromophytes, cryptophytes, photosynthetic dinoflagellates (which may include several distinct chloroplast lineages), and rhodophytes. Each may have undergone one or more separate chloroplast-forming symbiogenetic event. See L. Margulis et al., eds., 1990, *Handbook of Protoctista*, Boston: Jones and Bartlett. See also A. H. Knoll, 1992, The early evolution of eukaryotes: A geological perspective, *Science* 256:622–627.

4. Undulipodia are discused at length in L. Margulis and M. McMenamin, 1991, Kinetosome-Centriolar DNA: Significance for endosymbiosis theory, *Treballs de la Societat Catalana de Biologia* 41:5–16.

5. The full reference is R. H. Whittaker, 1959, On the broad classification of organisms, *Quarterly Review of Biology* 34:210–226. For a more recent analysis of the five kingdoms, see L. Margulis, 1993, *Symbiosis in Cell Evolution: Microbial Evolution in the Archean and Proterozoic Eons*, 2nd edition, New York: W. H. Freeman.

6. See H. J. Hofmann, 1985, Precambrian carbonaceous megafossils, in D. F. Toomey and M. H. Nitecki, eds., *Paleoalgology*, pp. 20–23, Berlin: Springer-Verlag.

7. For recent information on the supercontinent Rodinia, see C. M. Powell, Z. X. Li, M. W. McElhinny, J. G. Meert, and J. K. Park, 1993, Paleomagnetic constraints on the timing of the Neoproterozoic breakup of Rodinia and the Cambrian formation of Gondwana, *Geology* 21:889–892.

8. See M. McMenamin, 1993, Osmotrophy in fossil protoctists and early animals, *Invertebrate Reproduction and Development* 23:165–166.

9. See S. Conway Morris, 1993, Ediacaran-like fossils in Cambrian Burgess Shale–type faunas of North America, *Palaeontology* 36:593–635.

10. See W. Stolzenburg, 1990, When life got hard, *Science News* 138:120–123, and W. Stolzenburg, 1990, When, why did life suddenly become hard?, *Orlando Sentinel*, 18 November 1990, p. G-7.

11. See S. M. Stanley, 1973, An ecological theory for the sudden origin of multicellular life in the late Precambrian, *Proceedings of the National Academy of Sciences (USA)* 70:1486–1489, and R. Cowen, 1976, *History of Life*, New York: McGraw-Hill.

12. Cambrian foraminifera are discussed in S. J. Culver, 1991, Early Cambrian foraminifera from West Africa, *Science* 254:689–691.

4: Land Life Before Hypersea

1. See R. J. Horodyski and L. P. Knauth, 1994, Life on land in the Precambrian, *Science* 263:494–498, and L. P. Knauth and R. J. Horodyski, 1993, Evidence for and implications of life on land in the Proterozoic, *Geological Society of America Abstracts with Program* 26:A-80. For discussion of possible two-billion-

year-old actinomycete soil bacteria, see B. M. Waggoner, 1994, Fossil actinomycete in Eocene-Oligocene Dominican amber, *Journal of Paleontology* 68:398–401.

2. For information on rock varnish bacteria, see R. I. Dorn, 1991, Rock varnish, *American Scientist* 79:542–553.

3. The giant honey mushroom fungus is described in M. L. Smith, J. N. Bruhn, and J. B. Anderson, 1992, The fungus *Armillaria bulbosa* is among the largest and oldest living organisms, *Nature* 356:428–430.

4. For description of these early fungi, see M. A. Sherwood-Pike and J. Gray, 1985, Silurian fungal remains—probable records of the Class Ascomycetes, *Lethaia* 18:1–20.

5. The Jurassic shelf fungus *Phellinites* is described by R. Singer and S. Archangelsky, 1958, A petrified Basidiomycete from Patagonia, *American Journal of Botany* 45:194–198.

6. For more information on Antarctic lichens, see M. Bartusiak, 1983, Living in rock and lichen it, *Science 83* 4:74, and E. I. Friedmann and R. Ocampo-Friedmann, 1984, The Antarctic cryptoendolithic ecosystem: Relevance to exobiology, *Origins of Life* 14:771–776. If fossils occur on other planets in our solar system, they are likely to be in the form of fossils left by cryptoendolithic organisms; see E. I. Friedmann and R. Weed, 1987, Microbial trace-fossil formation, biogenous and abiotic weathering in the Antarctic cold desert, *Science* 236:703–705. For discussion of cryptoendolithic cyanobacteria and algae of the Colorado Plateau, see R. A. Bell and M. R. Sommerfeld, 1993, Cryptoendolithic algae of semiarid lands and deserts in the southwestern United States, *American Journal of Botany* 80:93.

7. Here is our ballpark calculation of potential cryptoendolithic biomass worldwide. Quartz sands have constituted between 2.7 and 8.6 percent of both sedimentary and volcanogenic rock laid down on continents throughout earth history. Averaging these numbers as a rough guide to the proportion of sedimentary rocks exposed on land that are sandstones, and cutting that average number in half to account for surface rocks that are neither sedimentary nor volcanogenic (such as granites and metamorphic rocks) and are not porous enough to harbor the tenacious lichens, and then multiplying by the global surface area of land (149 ×

10^6 km^2 at present) and multiplying the above value by 0.31 kg/ m^2 of lichen in sandstone, we conclude that there potentially could have been 130 trillion kilograms (gigatons) of lichen living in porous sandstone in an ancient world free of vascular plants. Of this, assume that 50 percent, or 65 gigatons, is dry weight organic matter. Of the dry weight organic matter, assume it is about 44 percent carbon atoms by weight. (The 44 percent is derived from the formula for the carbohydrate lichenin, obtained from Iceland moss and other lichens. Lichenin has the same empirical formula as starch, that is, $(C_6H_{10}O_5)_n$, with n being an integer.)

Calculating with these estimates, the total fixed carbon in a hypothetical "lichenland" is 29 gigatons.

Calculation of cryptoendolithic lichen biomass, extrapolated for land surface exposed sandstone, is as follows:

$$\frac{(2.7+8.6)}{2} \times \frac{1}{2} \times$$
$$(1.49 \times 10^6 \text{ km}^2) \times \frac{10^6 \text{m}^2}{1 \text{ km}^2} \times (0.31 \text{ kg/m}^2) = 130 \times 10^{12} \text{ kg}.$$

8. The Witwatersrand structures were originally described by D. K. Hallbauer and K. T. Van Warmelo, 1974, Fossilized plants in thucolite from Precambrian rocks of the Witwatersrand, South Africa, *Precambrian Research* 1:199–212. See also D. K. Hallbauer, H. M. Jahns, and H. A. Beltmann, 1977, Morphological and anatomical observations on some Precambrian plants from the Witwatersrand, South Africa, *Geologische Rundschau* 66:477–491. The interpretation in this work was challenged by P. Cloud and D. Pierce, 1977, Experimental production of pseudomicrofossils, *Geological Society of America Abstracts with Program* 9:102. The Alaskan gold bacterial fossils are described in J. R. Watterson, 1991, Preliminary evidence for the involvement of budding bacteria in the origin of Alaskan placer gold, *Geology* 20:315–318.

9. For these criteria for the recognition of the oldest vascular plants, see W. G. Chaloner, 1988, Early land plants—the saga of a great conquest, in W. Greuter and B. Zimmer, eds., *Proceedings of XIV International Botanical Congress*, 1988, Koeltz: Köigstein/ Taunus.

10. For interpretations of *Longfengshania*, see Zhang Z., 1988, *Long-fengshania* Du emend.: An earliest record of bryophyte-like fossils, *Acta Palaeontologica Sinica* 27:416–426; and H. J. Hofmann, 1985, Precambrian carbonaceous megafossils, in D. F. Toomey and M. H. Nitecki, eds., *Paleoalgology: Contemporary Research and Applications*, 1985, Berlin: Springer-Verlag, pp. 20–23.

11. The putative Proterozoic trilete spores are described in N. A. Volkova, 1976, O nakhodke Dokembriishikh spor s tetradnim rubtsom, *Paleontologiya Morskaya Geologiya Mezhdunarodny Geologicheskiya Kongressa, XXV Session*, Akademiya Nauk SSSR, Ministerstvo Geologichestvo SSSR, "Nauka," pp. 14–18.

12. Jane Gray and Arthur J. Boucot (1978, The advent of land plant life, *Geology* 6:489–492) tentatively interpreted these fossils with supposed trilete scares as an "independent, parallel development of a land plant–type character," and further suggested that such a terrestrial floral evolution failed in the Proterozoic, only to be followed by a successful colonization of land by vascular plants in the Ordovician. William A. Shear (1991, The early development of terrestrial ecosystems, *Nature* 351:283–289) considers this to still be a plausible suggestion, especially in light of arguments concerning the land plant nature of *Longfengshania* (Zhang Z., 1988, *Longfengshania* Du emend.: An earliest record of bryophyte-like fossils, *Acta Palaeontologica Sinica* 27:416–426).

5: Exploiting Osmosis for Hypersea Physiology

1. What might be called "anti-ecophysiology" is employed by D. C. Wight, 1987, Non-adaptive change in early land plant evolution, *Paleobiology* 13:208–214.

2. For more on ecophysiological considerations in the evolution of land life, see R. D. K. Thomas, 1984, When and how did plant and animals take to the land?, *Paleobiology* 10:1–8, and J. A. Raven, 1985, Comparative physiology of plant and arthropod land adaptation, *Philosophical Transactions of the Royal Society of London* 311B:273–288.

3. For information concerning land snails and osmoregulation, see C. Little, 1990, *The Terrestrial Invasion: An Ecophysiological Approach to the Origins of Land Animals*, Cambridge: Cambridge University Press, and A. Friday and D. S. Ingram, 1985, *The*

Cambridge Encyclopedia of Life Sciences, Cambridge: Cambridge University Press.

4. A technical discussion of the use of osmotic pressure (osmotic potential) and pressure potential by plants is provided by V. A. Greulach, 1973, *Plant Function and Structure*, New York: Macmillan.

5. For desiccation-resisting fats and oils, see A. P. Kelley, 1950, *Mycotrophy in Plants*, Waltham, Massachusetts: Chronica Botanica.

6. For introduction of the term *geophysiology*, see J. E. Lovelock, 1989, Geophysiology: The science of Gaia, *Reviews of Geophysics* 27:215–222.

7. The original paper on Gaia is J. E. Lovelock and L. Margulis, 1974, Atmospheric homeostasis by and for the biosphere: The Gaia hypothesis, *Tellus* 26:1–9.

8. A brief but good translation of one of Vernadsky's major articles has been available for some time as V. I. Vernadsky, 1945, The biosphere and the noösphere, *American Scientist* 33:1–12. See also A. H. Knoll, 1984, Review of *Traces of Bygone Biospheres*, *American Scientist* 72:200–201.

9. The Vernadsky's Law quotation is discussed in A. I. Perelman, 1979, *Geokhimiya*, Moscow: Vysshaya Shkola.

10. Vernadsky's derivation of biogeochemical constants is from V. I. Vernadsky, 1926, *Nastavleniya dla opredeleniya geokhimicheskikh postoyannykh. 2. Opredeleniye geochimicheskoi inergii (velichiny $\triangle,V,e$) nekotorykh grupp nasekomykh*, Leningrad: Akademiya Nauk. His derivation is:

$\triangle$ = optimal number of generations per day

k_1 = the greatest dimension (average value) of the organism in cm

V_1 = the velocity of bacteria

For bacteria, take $\triangle = 64$, $k_1 = 1$ micron $= .0001$ cm. Then:

$$V_1 = \frac{13963 \cdot \triangle}{18.71 - (\log_{10}k_1)}$$

$$V_1 = \frac{13963 \cdot (64)}{18.71 - (-4)}$$

$$V_1 = 39,349 \text{ cm/sec} = .393 \text{ km/sec}$$

$$(.393 \text{ km/sec})(5 \text{ sec})(.6214 \text{ miles/1 km}) = 1.22 \text{ miles}$$

Or, in other words, the velocity of bacteria on the surface of the planet works out to be about 1.22 miles in five seconds, assuming of course perfect survivorship of progeny and geometric rates of population increase (conditions which never actually occur in nature).

The velocity formula used in the above example can be explained and confirmed as follows. This velocity formula has two forms:

$$V_1 = \frac{13963 \cdot \Delta}{18.71 - (\log_{10}k_1)}, \quad \begin{array}{l} \Delta = \text{generations per day} \\ k_1 = \text{greatest dimension of creature} \end{array}$$

$$V_1 = \frac{13963 \cdot \Delta}{\log_{10} N_{max}}, \quad \Delta = \text{generations per day}$$

$$N_n = 2^{n\Delta} \quad \begin{array}{l} n = \text{number of days} \\ N_n = \text{number of individuals extant after} \\ \quad\quad n \text{ days of reproduction} \\ N_{max} = \text{maximum possible number of} \\ \quad\quad\quad \text{individuals of a given species} \\ \quad\quad\quad \text{living on Earth} \end{array}$$

The mean radius of the Earth is 6.37×106 meters, and the surface area is equal to 5.099×10^{14} m^2 or 5.099×10^{18} cm^2. The base ten logarithm of this last number equals 18.707. So, comparing the denominators of the two velocity formulas above,

$$18.707 - \log_{10}(k_1) = \log_{10} N_{max}$$

$$18.707 = \log_{10} N_{max} + \log_{10}(k_1)$$

$$18.707 = \log_{10}(N_{max})(k_1)$$

$$10^{18.707} = (k_1)(N_{max})$$

$$N_{max} = \frac{10^{18.707}}{k_1}$$

Or, to put it differently, the maximum number of creatures equals their average maximum dimension divided into the surface area of the earth. This truism demonstrates that the calculated value of V_1 is valid.

6: Hypermarine Aspects of Plants and Fungi

1. For description of the oldest accepted trilete spore tetrads, see J. Gray, 1993, *Geological Society of America Abstracts with Program* 25:A-81. See also J. Gray, D. Massa, and A. J. Boucot, 1982, Caradocian land plant microfossils from Libya, *Geology* 10:197–201.

2. For descriptions of cryptospores, see P. K. Strother, 1991, A classification schema for the cryptospores, *Palynology* 15:219–236; for nematoclast cuticular sheet descriptions, see P. G. Gensel, N. G. Johnson, and P. K. Strother, 1991, Early land plant debris (Hooker's 'Waifs and Strays'?), *PALAIOS* 5:520–547.

3. For discussion of fossil conducting cells, see P. Kenrick, D. Edwards, and R. C. Dales, 1991, Novel ultrastructure in water-conducting cells of the Lower Devonian plant *Sennicaulis hippocrepiformis*, *Palaeontology* 34:751–766.

4. Banks's dichotomy is outlined in H. P. Banks, 1968, The early history of land plants, in E. T. Drake, ed., *Evolution and Environment*, pp. 73–107, New Haven: Yale University Press.

5. *Baragwanathia* was described by W. H. Lang and I. C. Cookson, 1935, On a flora, including vascular land plants, associated with *Monograptus* in rocks of Silurian age, from Victoria, Australia, *Philosophical Transactions of the Royal Society of London* 224B:421–449.

6. For description of Irish *Cooksonia*, see D. Edwards and J. Feehan, 1980, Records of *Cooksonia*-type sporangia from late Wenlock strata in Ireland, *Nature* 287:41–42; for the problem concerning *Cooksonia*'s vascular system, see P. G. Gensel, 1991, Notes on the cuticular morphology of *Sawdonia acanthotheca*, particularly in regard to emergences, *Neues Jahrbuch für Geologie und Paläontologie. Abhandlungen* 183:49–59. For cooksoniid distribution, see H. P. Banks, 1975, Early vascular plants: Proof and conjecture, *BioScience* 25:730–727. Information regarding study of cooksoniid spores within coalified sporangia may be found in D. Edwards, 1993, New insights into early land vegetation, *Geological Society of America Abstracts with Program* 25: A-81–A-82.

7. For information on zosterophylls, see P. G. Gensel, 1991, Notes on the cuticular morphology of *Sawdonia acanthotheca*, particularly in regard to emergences, *Neues Jahrbuch für Geologie und Paläontologie. Abhandlungen* 183:49–59; A. G. Lyon and D. Edwards, 1991, The first zosterophyll from the Lower Devonian Rhynie Chert, Aberdeenshire, *Transactions of the Royal Society of Edinburgh* 82:323–332; and K. J. Niklas and H. P. Banks, 1990, A reevaluation of a zosterophyllophytina with comments on the origin of lycopods, *American Journal of Botany* 77:274–283.

8. For information concerning the Chinese Posongchong Formation, and discussion of novel character traits in early Devonian plants, see C. S. Li and D. Edwards, 1992, A new genus of early land plants with novel strobilar construction from the Lower Devonian Posonchong Formation, Yunnan Province, China, *Palaeontology* 35:257–272. The change to cosmopolitan floras in the late Devonian is discussed in H. D. Pflug and K. F. Prössl, 1991, Palynostratigraphical and paleobotanical studies in the pilot hole of the German continental deep drilling program: Results and implications, *Scientific Drilling* 2:13–33.

9. For discussion of bilateral symmetry in zosterophylls and their descendants, see P. Kenrick and D. Edwards, 1988, A new zosterophyll from a recently discovered exposure of the Lower Devonian Senni Beds in Dyfed, Wales, *Botanical Journal of the Linnean Society* 98:97–115.

10. See J. G. Duckett and R. Ligrone, 1992, A light and electron microscope study of the fungal endophytes in the sporophyte and gametophyte of *Lycopodium cernuum* with observations on the gametophyte-sporophyte junction, *Canadian Journal of Botany* 70:58–72.

11. Enation theory and telome theory are discussed at length in W. N. Stewart, 1983, *Paleobotany and the Evolution of Plants*, Cambridge: Cambridge University Press; and P. G. Gensel, 1991, Notes on the cuticular morphology of *Sawdonia acanthotheca*, particularly in regard to emergences, *Neues Jahrbuch für Geologie und Paläontologie. Abhandlungen* 183:49–59. See also P. M. Bonamo, H. P. Banks, and J. D. Greirson, 1988, *Leclercqia, Haskinsia*, and the role of leaves in delineation of Devonian lycopod genera, *Botanical Gazette* 149:222–239.

12. For information on the Canadian trimerophyte, see P. G. Gensel,

1984, A new Lower Devonian plant and the early evolution of leaves, *Nature* 309:785–787.

13. For discussion of the rotting lycophyte stem interiors, see T. Swain and G. Cooper-Driver, 1981, Biochemical evolution in early land plants, in K. J. Niklas, ed., *Paleobotany, Paleoecology, and Evolution*, pp. 103–134, New York: Praeger.

14. For discussion of *Amyelon iowense*, see A. A. Cridland, 1964, *Amyelon* in American coal-balls, *Palaeontology* 7:186–209.

15. Suggestions that early seeds were evolved directly from homosporous stock are dismissed by W. G. Chaloner and A. R. Hemsley, 1991, Heterospory: Cul-de-sac or pathway to the seed?, in S. Blackmore and S. H. Barnes, eds., *Pollen and Spores*, pp: 151–167, Oxford: Clarendon Press.

16. For caveats on the interpretation of heterospory, see T. N. Taylor and D. F. Brauer, 1983, Ultrastructural studies of in situ Devonian spores: *Barinophyton citrulliforme*, *American Journal of Botany* 70:106–112; R. B. Thompson, 1927, Evolution of the seed habit in plants, *Transactions of the Royal Society of Canada* 21:229–272; and R. B. Thompson, 1934, Heterothally and the seed habit versus heterospory, *New Phytologist* 33:41–44.

17. For the report of megaspores and microspores in coal balls, see F. D. Reed, 1926, Flora of an Illinois coal ball, *Botanical Gazette* 81:460–469. The retirement tribute to Fredda Reed is by B. F. Thomson (1960, Fredda D. Reed—Professor of Plant Science, *Mount Holyoke Alumnae Quarterly* 42:61).

18. For discussion of heterospory in *Barinophyton*, see T. N. Taylor and D. F. Brauer, 1983, Ultrastructural studies of in situ Devonian spores: *Barinophyton citrulliforme*, *American Journal of Botany* 70:106–112.

19. For discussion of the diversity of megaspores and *Cystosporites devonicus*, see M. G. Mortimer and W. G. Chaloner, 1967, Devonian megaspores from the Wyboston Borehole, Bedfordshire, England, *Palaeontology* 10:189–213; and W. G. Chaloner and J. M. Pettitt, 1964, A seed megaspore from the Devonian of Canada, *Palaeontology* 7:29–36.

20. Early fine rhizome traces in fossil soil are discussed in S. G. Driese, D. E. Fastovsky, C. I. Mora, and P. G. Gensel, 1993, Sedimentologic, paleontologic, and stable isotopic evidence for Late Silurian land plants in central Pennsylvania, *Geological Society of America Abstracts with Program* 25:A-81.

21. For more details on the case of the fossil "serpent," see D. Ager, 1993, *The New Catastrophism*, Cambridge: University Press, pp. 8–9.

22. For discussion of air conduction in plant roots, see R. A. Spicer, 1989, Physiological characteristics of land plants in relation to environment through time, *Transactions of the Royal Society of Edinburgh, Earth Science* 80:321–329; I. A. Mendelssohn, K. L. McKee, and W. H. Patrick, 1981, Oxygen deficiency in *Spartina alterniflora* roots: Metabolic adaptation to anoxia, *Science* 214:439–441; and M. D. Bertness, 1992, The ecology of a New England salt marsh, *American Scientist* 80:260–268.

23. For discussion of upland cordaites, see R. L. Leary, 1981, Early Pennsylvanian geology and paleobotany of the Rock Island County, Illinois, area. Part 1: Geology, *Illinois State Museum, Reports of Investigations* 37:1–88.

24. For information on the earliest wind-dispersed seeds, see N. P. Rowe, 1992, Winged Late Devonian seeds, *Nature* 359:682.

25. For description of *Tetrapterites,* see H. J. Sullivan and A. F. Hibbert, 1964, *Tetrapterites visensis*—a new spore-bearing structure from the Lower Carboniferous, *Palaeontology* 7:64–71.

26. For description of *Triletes,* see K. L. Alvin, 1966, Two cristate megaspores from the Lower Carboniferous of Scotland, *Palaeontology* 9:488–491.

27. For the *Neuropteris*-pteridosperm connection, see P. G. Gensel, 1988, On *Neuropteris* Brongniart and *Cardiopteridium* Nathorst from the Early Carboniferous Price Formation, southwestern Virginia, USA, *Review of Palaeobotany and Palynology* 54:105–119.

28. See H. M. Stoneley, 1956, *Hiltonia,* a new plant genus from the Upper Permian of England, *Annual Magazine of Natural History, Series 12* 9:713–720.

29. For a discussion of endosperm evolution, see W. E. Friedman, 1992, Evidence of a pre-angiosperm origin of endosperm— implications for the evolution of flowering plants, *Science* 225:336–339.

30. Concerning Bower's hypothesis and its alternative, see F. O. Bower, 1908, *The Origins of a Land Flora*, London: MacMillan; L. E. Graham, 1985, The origin of the life cycle of land plants, *American Scientist* 73:178–186; L. E. Graham and L. W. Wilcox, 1983, The occurrence and phylogenetic significance of puta-

tive placental transfer cells in the green alga *Coleochaete*, *American Journal of Botany* 70:113–120; and L. Celakovsky, 1874, *Bedeutung des Generationswechsels der Pflanzen*, Prague.

31. Concerning *Coleochaete*, see L. E. Graham, 1985, The origin of the life cycle of land plants, *American Scientist* 73:178–186; L. E. Graham and L. W. Wilcox, 1983, The occurrence and phylogenetic significance of putative placental transfer cells in the green alga *Coleochaete*, *American Journal of Botany* 70:113–120; and G. L. Stebbins and G. J. C. Hill, 1980, Did multicellular plants invade the land?, *American Naturalist* 115:342–353.

32. See P. A. Selden and D. Edwards, 1989, Colonisation of the land, in K. C. Allen and D. E. G. Briggs, eds., *Evolution and the Fossil Record*, pp. 122–152, London: Belhaven Press.

33. This discussion of *Lycopodium cernuum* was adapted from O. Degener, 1984, *Plants of Hawaii National Park Illustrative of Plants and Customs of the South Seas*, Ann Arbor, Michigan: Braun-Brumfield. See also D. P. Whittier, 1977, Gametophytes of *Lycopodium obscurum* as grown in axenic culture, *Canadian Journal of Botany* 55:563–567.

34. J. G. Duckett and R. Ligrone, 1992, A light and electron microscope study of the fungal endophytes in the sporophyte and gametophyte of *Lycopodium cernuum* with observations on the gametophyte-sporophyte junction, *Canadian Journal of Botany* 70:58–72.

35. Spongiophytes are discussed in P. G. Gensel, N. G. Johnson, and P. K. Strother, 1991, Early land plant debris (Hooker's 'Waifs and Strays'?), *PALAIOS* 5:520–547; and V. Krassilov, 1981, *Orestovia* and the origin of vascular plants, *Lethaia* 14:235–250. A lichen affinity for *Spongiophyton* is hypothesized by W. E. Stein, G. D. Harmon, and F. M. Hueber, 1993, *Spongiophyton* from the Lower Devonian of North America reinterpreted as a lichen, *American Journal of Botany* 80:93. That hypothesis is supported in W. E. Stein, G. D. Harmon, and F. M. Hueber, 1993, Lichens in the Lower Devonian of North America, *Geological Society of America Abstracts with Program* 25:A-82.

36. See W. E. Stein, G. D. Harmon, and F. M. Hueber, 1993, *Spongiophyton* from the Lower Devonian of North America reinterpreted as a lichen, *American Journal of Botany* 80:93, and

W. E. Stein, G. D. Harmon, and F. M. Hueber, 1993, Lichens in the Lower Devonian of North America, *Geological Society of America Abstracts with Program* 25:A-82.

37. For discussion of nematophytes, see R. Schmid, 1976, Septal pores in *Prototaxites*, an enigmatic Devonian plant, *Science* 191:287–288; and D. Edwards, 1980, Early land floras, in A. L. Panchen, ed., *The Terrestrial Environment and the Origin of Land Vertebrates*, pp. 55–85, London: Academic Press.

38. For studies of *Oxyria digyna*, see H. A. Mooney and W. D. Billings, 1961, Comparative physiological ecology of arctic and alpine populations of *Oxyria digyna*, *Ecological Monographs* 31:1–29.

39. See W. D. Billings, 1965, *Plants and the Ecosystem*, Belmont, California: Wadsworth. See also W. D. Billings, 1970, *Plants, Man and the Ecosystem*, Belmont, California: Wadsworth.

40. *Sporogonites* is described by H. N. Andrews, 1960, Notes on Belgian specimens of *Sporogonites*, *Palaeobotanist* 7:85–89.

41. Following is the calculation of the number of putative conductive tubes in the trunk of *Prototaxites*. Since, for cross-sectional area, $A = \pi r^2$, the cross-sectional area of the trunk is $A \cong 3.14 \times 0.5 m^2 = 0.786 m^2$. The tubes are 30 percent of the total cross-sectional area of the trunk, or $0.786 m^2 \times .3 = 0.236 m^2$. The cross-sectional area of a single tube a $\cong 3.14 \times 17.5 \mu^2 = 9.62 \times 10^2 \mu^2$. Note that $1 m^2 = 1 \times 10^{12} \mu^2$, since a micron is a thousandth of a millimeter and a millimeter is a thousandth of a meter. So $0.236 m^2 = 0.236 \times 10^{12} \mu^2$. By simple division, the total number of tubes running through the trunk is:

$$\frac{(.236 \times 10^{12} \mu^2)}{(9.62 \times 10^2 \mu^2)} = 2.45 \times 10^8 = 245 \text{ million}$$

42. For transpiration rate of tropical trees, see C. F. Jordan and J. R. Kline, 1977, Transpiration of trees in a tropical rainforest, *Journal of Applied Ecology* 14:853–860. The transpiration equation derived in this paper is t = 0.73s - 31, where t is the transpiration rate in liters per day, and s is the calculated sapwood area of trees in square centimeters. In the trees measured, the ratio of sapwood cross-sectional area to total cross-sectional area of the tree (sapwood area plus heartwood area) was quite variable, ranging from 0.13 to 0.78. An average of thirteen ratios in

which the transition from heartwood to sapwood was distinct gives an average ratio of 0.388. In other words, the sapwood in the average tree of the thirteen has 38.8 percent of its total cross-sectional area composed of sapwood.

So, for a Venezuelan tree 1 m in diameter, the cross-sectional area is 0.786 m^2 or 7,860 cm^2. Of this,

$$7,860 \text{cm}^2 \times 0.388 = 3,050 \text{cm}^2$$

is sapwood. A tree with this much sapwood transpires, according to the linear formula of Jordan and Kline (1977):

$$t = 0.73 \times 3050 \text{cm}^2 - 31 = 2,196 \text{ liters per day.}$$

Since *Oxyria* transpires 0.03 g/cm^{2}hr, and 1 g water occupies 1 cc of volume, scaling up the *Oxyria* transpiration rate to liters gives:

$$(0.03 \text{ cc/cm}^2\text{hr}) \, (1/1000 \text{ liters/cc}) \, (24 \text{ hr/day}) =$$
$$7.2 \times 10^{-4} \text{ l/cm}^2\text{day.}$$

Thus, the transpiring *Prototaxites* trunk is estimated here to be able to support a photosynthetic organ with a surface area of:

$$(2,196 \text{ liters/day}) \, / \, (7.2 \times 10^{-4} \text{ liters/cm}^2\text{day}) =$$
$$3.05 \times 10^6 \text{ cm}^2 = 305 \text{ m}^2$$

43. Rhynie gametophytes are discussed by W. Remy and R. Remy, 1980, Devonian gametophytes with anatomically preserved gametangia, *Science* 208:295–296.

44. For discussion of the Rhynie fossil plants, see A. G. Lyon and D. Edwards, 1991, The first zosterophyll from the lower Devonian Rhynie Chert, Aberdeenshire, *Transactions of the Royal Society of Edinburgh* 82:323–332; E. W. Berry, 1945, The beginnings and history of land plants, *Johns Hopkins University Studies in Geology* 14:9–91; and V. Krassilov, 1981, *Orestovia* and the origin of vascular plants, *Lethaia* 14:235–250.

45. The land plants with a complex level of stem-leaf differentiation are described by P. G. Gensel, 1984, A new Lower Devonian plant and the early evolution of leaves, *Nature* 309:785–787.

46. For discussion of the fungus-covered arthropod, see W. D. I. Rolfe, 1980, Early invertebrate terrestrial faunas, in A. L.

Panchen, ed., *The Terrestrial Environment and the Origin of Land Vertebrates*, pp. 117–157, London: Academic Press.

47. For discussion of internal proliferation of mycorrhizal spores, see R. E. Koske, 1984, Spores of VAM fungi inside spores of VAM fungi, *Mycologia* 76:853–862. K. A. Pirozynski and Y. Dalpé (1989, Geological history of the Glomaceae with particular reference to mycorrhizal symbiosis, *Symbiosis* 7:1–36) identified *Glomus* in the Rhynie Chert.

48. For discussion of the Antarctic arbuscles, see S. P. Stubblefield, T. N. Taylor, and J. M. Trappe, 1987, Fossil mycorrhizae: A case for symbiosis, *Science* 237:59–60.

49. The *Pinus* mycorrhizal experiments were reported in J. L. Harley, 1956, The mycorrhiza of forest trees, *Endeavor* 15:43–48.

50. See S. M. Schwab, J. A. Menge, and R. T. Leonard, 1983, Comparison of stages of vesicular-arbuscular mycorrhiza formation in sudangrass grown at two levels of phosphorus nutrition, *American Journal of Botany* 70:1225–1232.

51. For more information on the relationship between orchids and their mycorrhizal fungi, see I. Tribe, 1971, *The Plant Kingdom*, New York: Bantam. See also H. N. Rasmussen, 1992, Seed dormancy patterns in *Epipactis palustris* (Orchidaceae), requirements for germination and establishment of mycorrhiza, *Physiologia Plantarum* 86:161–167.

52. The fungal fusion hypothesis was proposed by K. A. Pirozynski and D. W. Malloch, 1975, The origin of land plants: A matter of mycotrophism, *BioSystems* 6:153–164. The fungus proposed by them to have fused with a green alga to form plants is actually an oomycete, which is not a true fungus because it forms swimming propagules. No true fungi have this capability.

53. For more on the fungal fusion hypothesis, see P. R. Atsatt, 1991, Fungi and the origin of land plants, in L. Margulis and R. Fester, eds., *Symbiosis as a Source of Evolutionary Innovation*, Cambridge, Massachusetts: MIT Press. See also R. Jorgensen, 1993, The origin of land plants: A union of alga and fungus advanced by flavonoids?, *BioSystems* 31:193–207.

54. Ironically, many plants, and even other fungi, use chitinases as a chemical defense against pathogenic fungi; see C. Kunz, A. Ludwig, and Y. Bertheau, 1992, Evaluation of the antifungal activity

of the purified chitinase-1 from the filamentous fungus *Aphanocladium album*, *FEMS Microbiology Letters* 90:105–109. See also Y. Tirilly, F. Lambert, and D. Thouvenot, 1991, Bioproduction of [C-14] deoxyphomenone, a fungistatic metabolite of the hyperparasite *Dicyma pulvinata*, *Phytochemistry* 30:3963–3965.

55. See R. Jorgensen, 1993, The origin of land plants: A union of alga and fungus advanced by flavonoids?, *BioSystems* 31:193–207.

56. For the fossil record of the charophytes, see V. P. Maslov and V. A. Vakhrameev, eds., 1966, *Iskopaemye Kharofity SSSR*, Moscow: Iztadel'stvo "Nauka." For beautiful scanning electron micrographs of fossil charophytes, see M. Feist and N. Grambast-Fessard, 1984, New Porocharaceae from the Bathonian of Europe: Phylogeny and paleoecology, *Palaeontology* 27:295–305.

57. The parasitized Rhynie *Palaeonitella* is described in T. N. Taylor, W. Remy, and H. Hass, 1992, Parasitism in a 400-million-year-old green alga, *Nature* 357:493–494; and T. N. Taylor, H. Hass, and W. Remy, 1992, Devonian fungi: Interactions with the green alga *Palaeonitella*, *Mycologia* 84:901–910.

58. Discussion of the relationship between tall fescue and its endophyte may be found in D. M. Ball, J. F. Pedersen, and G. D. Lacefield, 1993, The Tall-Fescue endophyte, *American Scientist* 81:370–379; D. Ball, S. Schmidt, G. Lacefield, C. Hoveland, W. C. Young, 1991, *Tall Fescue/Endophyte/Animal Relationships*, Salem, Oregon: Oregon Tall Fescue Commission; and D. Ball, M. Putnam, G. Lacefield, and C. Hoveland, 1991, *Fescue for Horses: Problem or Opportunity?*, Salem, Oregon: Oregon Tall Fescue Commission.

59. Fungal endophytes are good news for those who maintain lawns. New endophyte-bearing strains such as the perennial ryegrasses "Advent," "Prelude II," "Pleasure," and "Riviera," the fine fescues "Reliant" and "Warwick," and the tall fescue varieties "Titan" and "Shenandoah" stay healthier on their own and require fewer pesticides. These grasses are most easily established by planting seed innoculated with the endophyte. See Anonymous, 1993, Solving problems—friendly fungus aids lawns, *Fine Gardening*, July/August:24.

60. See R. Ligrone, K. Pocock, and J. G. Duckett, 1993, A compar-

ative ultrastructural study of endophytic basidiomycetes in the parasitic achlorophyllous hepatic *Cryptothallus mirabilis* and the closely allied photosynthetic species *Aneura pinguis* (Metzgeriales), *Canadian Journal of Botany* 71:666–679.

7: Animals and Hypersea

1. The Lobopodia may be a "grab bag" group containing both uniramous and biramous forms. See H. Gee, 1993, Paleontology and existential angst, *Nature* 364:674. See also G. Budd, 1993, A Cambrian gilled lobopod from Greenland, *Nature* 364:709–711.

2. For discussion of *Microdictyon* and *Xenusion,* see L. Ramsköld and Hou Xianguang, 1991, New early Cambrian animal and onychophoran affinities of enigmatic metazoans, *Nature* 351:225–228.

3. For discussion of Gilboa and Rhynie arthropods, see W. A. Shear, 1993, One small step for an arthropod, *Natural History* 102:47–51. See also W. A. Shear, 1993, The composition of Silurian-Devonian terrestrial arthropod faunas, *Geological Society of America Abstracts with Program* 25:A-81.

4. For discussion of arthropleurid trace fossils, see D. E. G. Briggs, W. D. I. Rolfe, and J. Brannan, 1979, A giant myriapod trail from the Namurian of Arran, Scotland, *Palaeontology* 22:273–291. See also D. E. G. Briggs, A. G. Plint, and R. K. Pickerill, 1984, *Arthropleura* trails from the Westphalian of eastern Canada, *Palaeontology* 27:843–855.

5. For discussion of *Amynilyspes,* see J. T. Hannibal and R. M. Feldmann, 1981, Systematics and functional morphology of oniscomorph millipedes (Arthropoda, Diplopoda) from the Carboniferous of North America, *Journal of Paleontology* 55:730–746.

6. Regarding early scorpions such as *Praearcturus,* see A. J. Jeram, 1993, Scorpions from the Middle Devonian of New York, and their implications for morphological disparity in early terrestrial scorpions, *Geological Society of America Abstracts with Program* 25:A-83.

7. For discussion of animal and fungal detoxification of plant matter, see W. A. Shear, 1991, The early development of terrestrial ecosystems, *Nature* 351:283–289. See also I. K. Ross, 1979, *Biology of the Fungi,* New York: McGraw-Hill. The controversy regarding the Litter Box hypothesis is developed in R. Beer-

bower, 1993, Ends and beginnings: Early tracheophyte evolution, *Geological Society of America Abstracts with Program* 25: A-82; P. A. Selden, 1993, Plant-animal interactions in early terrestrial ecosystems, *Geological Society of America Abstracts with Program* 25:A-82; and H. P. Banks and B. J. Colthart, 1993, Plant-animal-fungal interactions in Early Devonian trimerophytes from Gaspé, Canada, *American Journal of Botany* 80:992–1001.

8. Wounds in Devonian plants are described in C. A. Trant and P. G. Gensel, 1985, Branching in *Psilophyton:* A new species from the Lower Devonian of New Brunswick, Canada, *American Journal of Botany* 72:1256–1273.

9. See W. A. Shear, 1991, The early development of terrestrial ecosystems, *Nature* 351:283–289.

10. See P. A. Selden, 1993, Plant-animal interactions in early terrestrial ecosystems, *Geological Society of America Abstracts with Program* 25:A-82.

11. See R. Beerbower, 1993, Ends and beginnings: Early tracheophyte evolution, *Geological Society of America Abstracts with Program* 25:A-82.

12. For more on the pentastome-branchiurid link, see V. Pearse, J. Pearse, M. Buchsbaum, and R. Buchsbaum, 1987, *Living Invertebrates*, Palo Alto, California: Blackwell Scientific.

13. For discussion of putative Ordovician pentastomes, see D. Andres, 1989, Phospatisierte Fossilien aus dem unteren Ordoviz von Südschweden, *Berliner geowissenschaftliche Abhandlungen, Reihe A: Geologie und Palaeontologie* 106:9–19.

8: Parasitism in Hypersea

1. See N. E. Stork and C. H. C. Lyal, 1993, Extinction or 'co-extinction' rates?, *Nature* 366:307.

2. For discussion of the link between parasites and host sexuality, see W. D. Hamilton, R. Axelrod, and R. Tanese, 1990, Sexual reproduction as an adaptation to resist parasites: A review, *Proceedings of the National Academy of Sciences (USA)* 87:3566–3573. See also B. Holmes, 1993, Evolution's neglected superstars, *New Scientist* 140:30–33.

3. For discussion of *Leucochloridium* and *Dicrocoelium*, see M. Love, 1980, The alien strategy, *Natural History* 89:30–32.

4. See B. Holmes, 1993, Evolution's neglected superstars, *New Scientist* 140:30–33.

5. See M. Love, 1980, The alien strategy, *Natural History* 89:30–32.

6. Parasitism as the route to land for fungi is the suggestion of P. G. Kevan, W. G. Chaloner, and D. B. O. Savile, 1975, Interrelationships of early terrestrial arthropods and plants, *Palaeontology* 18:391–417. The relationship between the lichen mycobiont and phycobiont has been interpreted as controlled parasitism by V. Ahmadjian, 1982, The nature of lichens, *Natural History* 91:31–37.

7. *Baryonyx* is described in A. J. Charig and A. C. Milner, 1986, *Baryonyx*, a remarkable new therapod dinosaur, *Nature* 324:359–361.

8. The tragic case of Roberta Hanson is described in B. H. Kean, 1990, *M. D.: One Doctor's Adventures among the Famous and Infamous from the Jungles of Panama to a Park Avenue Practice,* New York: Ballantine.

9. For discussion of parasite survival during host extinction, see S. Conway Morris, 1990, Palaeoecology. 4.14 Parasitism, in D. E. G. Briggs and P. R. Crowther, eds., *Palaeobiology: A Synthesis*, pp. 179–181, Oxford: Blackwell.

10. For discussion of epiphytes of vascular plants, see D. H. Benzig, 1990, *Vascular Epiphytes*, Cambridge: Cambridge University Press.

11. For information on Mesozoic and living chalcids, see C. M. Yoshimoto, 1975, Cretaceous chalcidoid fossils from Canadian amber, *Canadian Entomologist* 107:499–528. See also E. E. Grissell and M. E. Schauff, 1990, *A Handbook of the Families of Nearctic Chalcidoidea (Hymenoptera)*, Washington, D. C.: Entomological Society of Washington.

12. For discussion of *Hymenaea*, see J. H. Langenheim, 1990, Plant resins, *American Scientist* 78:16–24; for a discussion of ancient leaf-spotting fungi (order Hemisphaeriales, family Hemisphaeriaceae), see V. Krassilov, 1981, *Orestovia* and the origin of vascular plants, *Lethaia* 14:235–250.

13. For discussion of *Xenocoeloma*, see H. Kondo, ed., 1972, *The Illustrated Encyclopedia of the Animal Kingdom, Volume 13*, New York: Danbury Press.

14. For discussion of ellobiopsids, see H. C. Whisler, 1990, Incertae sedis—Ellobiopsida, in L. Margulis et al., eds., *Handbook of Protoctista*, pp. 715–719, Boston: Jones and Bartlett. See also L. Margulis, H. I. McKhann, and L. Olendzenski, eds., 1993, *Illustrated Glossary of Protoctista*, Boston: Jones and Bartlett.

15. Regarding the parasite vectors hypothesis, a possible bias, involving the higher productivity of terrestrial ecosystems, may need to be compensated for, although this bias may already be partially compensated for by the fact that the oceanic carbon reservoir is approximately 20 times the size of the terrestrial carbon reservoir.

16. For discussion of food chain length, see F. Briand and J. E. Cohen, 1987, Environmental correlates of food chain length, *Science* 238:956–960.

17. For discussion of virulence, see P. W. Ewald, 1993, The evolution of virulence, *Scientific American* 268:86–88, and P. W. Ewald, 1994, *Evolution of Infectious Disease*, Oxford: Oxford University Press.

18. See L. Owens, 1993, Prevalence of *Cabirops orbionei* (Epicaridea, Cryptoniscidae) in Northern Australia—a biocontrol agent for bopyrids, *Australian Journal of Marine and Freshwater Research* 44:381–387.

9: Hypermarine Oceanography

1. Quotation from p. 55 of E. W. Berry, 1945, The beginnings and history of land plants, *Johns Hopkins University Studies in Geology* 14:9–91.

2. For discussion of sponges on mangrove roots, see A. M. Ellison and E. J. Farnsworth, 1992, The ecology of Belizean mangrove root-fouling communities: Patterns of epibiont distribution and abundance, and effects on root growth, *Hydrobiologia* 247:87–98.

3. For discussion of *Rhizosolenia*, see T. L. Hayward, 1993, The rise and fall of *Rhizosolenia*, *Nature* 363:709–712. See also T. A. Villareal, M. A. Altabet, and K. Culver-Rymsza, 1993, Nitrogen transport by vertically migrating diatom mats in the North Pacific Ocean, *Nature* 363:709–712.

4. Hypermarine upwelling is discussed further in M. A. S.

McMenamin and D. L. S. McMenamin, 1993, Hypersea and the Land Ecosystem, *BioSystems* 31:145–153.

5. For discussion of carnivorous plants, see B. E. Juniper, R. J. Robins, and D. M. Joel, 1989, *The Carnivorous Plants*, London and San Diego: Academic Press.

6. For discussion of *Monotropa*, see I. K. Ross, 1979, *Biology of the Fungi*, New York: McGraw-Hill.

7. The natural grafting in Marjorie Root's garden is reported in J. Danko, A surprise crops up for devoted gardener, *Daily Hampshire Gazette*, 2 August 1993, p. 12.

8. A stand of quaking aspen has been nominated as the world's largest clone. See M. C. Grant, 1993, The trembling giant: The quaking aspen, one of this country's most beautiful trees, also makes up the world's most massive organism, *Discover* 14:82–89. See also H. E. McLean, 1993, Romancing the clone, *American Forests* 99:8.

9. See J. N. Couch, 1938, *The Genus Septobasidium*, Chapel Hill: University of North Carolina Press.

10. For mention of injection of *Septobasidium* metabolites, see K. A. Pirozynski and D. L. Hawksworthy, 1988, Coevolution of fungi with plants and animals: Introduction and overview, in K. A. Pirozynski and D. L. Hawksworthy, eds., *Coevolution of Fungi with Plants and Animals*, pp. 1–29, London: Academic Press.

11. See A. E. Steele, 1984, Nematode parasites of sugar beet, in W. R. Nickle, ed., *Plant and Insect Nematodes*, pp. 507–569, New York: Marcel Dekker.

12. See H.-B. Jansson and G. O. Poinar, Jr., 1986, Some possible fossil nematophagous fungi, *Transactions of the British Mycological Society* 87:471–474. See also G. O. Poinar, Jr., 1992, *Life in Amber*, Stanford, California: Stanford University Press.

13. For the daisy-chain effect see D. J. Read, 1989, Mycorrhizas and nutrient cycling in sand dune ecosystems, *Proceedings of the Royal Society of Edinburgh*, B96:89–110.

14. Robinson argues that the Paleozoic coal accumulations resulted from evolutionary bottlenecks in the ability of early fungi to break down so much organic matter. See J. M. Robinson, 1991, Phanerozoic atmospheric reconstructions: A terrestrial perspective, *Palaeogeography, Palaeoclimatology, Palaeoecology (Global*

and Planetary Change Section) 97:51–62. The copious remains of fungi in Paleozoic coal are discussed in R. B. Wilson, 1993, Reassessment of the evidence for primary fusinite and degrado-fusinite, *Organic Geochemistry* 20:209–221.

15. For discussion of manna, see W. Gunther Plaut, 1981, *The Torah: A Modern Commentary*, New York: Union of American Hebrew Congregations.

16. See K. Esau, 1961, *Plants, Viruses, and Insects*, Cambridge, Massachusetts: Harvard University Press.

17. See K. Esau, 1961, *Plants, Viruses, and Insects*, Cambridge, Massachusetts: Harvard University Press.

18. See I. Tribe, 1971, *The Plant Kingdom*, New York: Bantam.

19. Pyotr A. Kropotkin's main work is *Mutual Aid: A Factor in Evolution*, reprinted in 1987 by Freedom Press (London).

20. For discussion of noösphere, see V. I. Vernadsky, 1945, The biosphere and the noösphere, *American Scientist* 33:1–12. Contrary to the assumptions of many scientists (e.g., A. H. Knoll, 1984, Review of *Traces of Bygone Biospheres, American Scientist* 72:200–201), the concept of the noösphere is a scientific one that need not be associated with teleological notions of purposeful evolution. For further consideration of this subject, see V. E. Ermolaeva, 1989, *Uchenie V. I. Vernadskogo o noosfere: Nauchno-analiticheskii obzor*, Moscow: Akademiia nauk SSSR.

21. Hypovirulence induced by dsRNA in Chestnut Blight Fungus is discussed by D. L. Nuss, 1992, Biological control of Chestnut Blight—an example of virus-mediated attenuation of fungal pathogenesis, *Microbiological Reviews* 56:561–576; G. H. Choi and D. L. Nuss, 1992, Hypovirulence of Chestnut Blight fungus conferred by an infectious viral cDNA, *Science* 257:800–803; and J. R. Newhouse and W. L. MacDonald, 1991, The ultrastructure of hyphal anastomoses between vegetatively compatible and incompatible virulent and hypovirulent strains of *Cryphonectria parasitica, Canadian Journal of Botany* 69:602–614.

10: Humans as Reservoirs of Hypersea

1. For analysis of the pork tapeworm incident, see P. M. Schantz et al., 1992, Neurocysticercosis in an Orthodox Jewish community

in New York City, *New England Journal of Medicine* 327:692–695.

2. For recent studies on malaria, see A. W. Taylor-Robinson, R. Stephen Phillips, A. Severn, S. Moncada, Foo Y. Liew, 1993, The role of T_H1 and T_H2 cells in a rodent malarial infection, *Science* 260:1931–1934.

3. For discussion of *Demodex*, see T. C. Cheng, 1986, *General Parasitology*, 2nd edition, Orlando, Florida: Academic Press.

4. For discussion of schistosomiasis, see S. R. Smithers and R. J. Terry, 1969, The immunology of schistosomiasis, *Advances in Parasitology* 7:41–93; S. R. Smithers and R. J. Terry, 1976, The immunology of schistosomiasis, *Advances in Parasitology* 14:399–422; and D. J. Hockley, 1973, Ultrastructure of the tegument of *Schistosoma*, *Advances in Parasitology* 11:233–305.

5. For discussion of parasite density, see C. H. Willey, 1941, The life history and bionomics of the trematode, *Zygocotyle lunata* (Paramphistomidae), *Zoologica* 26:65–88. See also T. C. Cheng, 1986, *General Parasitology*, 2nd edition, Orlando, Florida: Academic Press.

6. For discussion of *Nosema*, see P. F. Lai and E. U. Canning, 1980, Infectivity of a microsporidium of mosquitos (*Nosema algerae*) to larval stages of *Schistosoma mansoni* in *Biomphalaria glabrata*, *International Journal of Parasitology* 10:293–301.

7. For discussion of *Toxoplasma*, see L. D. Sibley and J. C. Boothroyd, 1992, Virulent strains of *Toxoplasma gondii* comprise a single clonal lineage, *Nature* 359:82–85.

8. For discussion of Marburg virus, see M. P. Kiley, N. J. Cox, L. H. Elliot, A. Sanches, R. Defries, M. J. Buchmeier, D. D. Richman, and J. B. McCormick, 1988, Physicochemical properties of Marburg virus—evidence for 3 distinct virus strains and their relationship to ebola virus, *Journal of General Virology* 69:1957–1967. See also E. D. Kilbourne, 1983, Are new diseases really new?, *Natural History* 92:28–32.

9. Most researchers think that HIV first evolved in nonhuman primates, and only recently made the jump to humans. On the other hand, Paul Ewald (1994, *Evolution of Infectious Diseases*, Oxford: Oxford University Press) argues that the virus has been present in humans for thousands of years, but has only now

become a deadly disease because of the modern upsurge in the typical number of sexual partners. This increase, he says, has led directly to increased virulence of HIV.

10. For discussion of AIDS and Idjwi sexual practices, see G. W. Shannon, 1991, AIDS: A search for origins, in R. Ulack and W. F. Skinner, eds., *AIDS and the Social Sciences—Common Threads, pp.8-29,* Lexington: University Press of Kentucky.

11. For discussion of viroids, see G. Epps, 1981, Viroids among us, *Science 81* 2:70−75.

12. See K. Esau, 1961, *Plants, Viruses, and Insects,* Cambridge, Massachusetts: Harvard University Press.

13. For discussion of the etiology of AIDS, see M. Nowak, R. M. Anderson, and A. R. McLean, 1991, Antigenic diversity thresholds and the development of AIDS, *Science* 254:963−969.

14. *Demodex* symptoms in AIDS patients are studied by J. P. de Jaureguiberry, F. Carsuzaa, C. Pierre, D. Arnoux, and D. Jaubert, 1993, Folliculite a *Demodex:* Une cause de prurit au cours de l'infection par le virus de l'immunodeficience humaine, *Annales de Medecine Interne Paris* 144:63−64.

15. For discussion of attempts to create an AIDS vaccine, see M. D. Daniel, F. Kirchhoff, S. C. Czajak, K. Sehgal, and R. C. Desrosiers, 1992, Protective effects of a live attenuated SIV vaccine with a deletion in the nef gene, *Science* 258:1938−1941.

16. For information on virus transformations in bacteria, see C. Schleper, K. Kubo, and W. Zillig, 1992, The particle SSV1 from the extremely thermophilic archaeon sulfolobus is a virus— demonstration of infectivity and of transfection with viral-DNA, *Proceedings of the National Academy of Sciences (USA)* 89:7645− 7649.

17. For more information on the evolutionary link between mycor-rhizae and plant diversification, see L. Simon, J. Bousquet, R. C. Lévesque, and M. Lalonde, 1993, Origin and diversifica-tion of endomycorrhizal fungi and coincidence with vascular land plants, *Nature* 363:67−69.

18. For discussion of *Rafflesia,* see G. Ismail, 1988, Conservation of the giant *Rafflesia* in Sabah, Malaysia, *Trends in Ecology and Evolution* 3:316−318.

abiogenic   Formed by physical, as opposed to biological, processes.

abiotic   Caused by physical, as opposed to biological, processes.

acritarch   Any small, roughly spherical fossil with a wall composed of a resistant organic material; usually removed from enclosing sediments by maceration (dissolving rock matrix with acid).

aerobe   An organism adapted to high levels of ambient oxygen. Cf. *anaerobe*.

alga   Any aquatic, eukaryotic, photosynthetic organism within the kingdom Protoctista; may be multicellular or unicellular. An alga is not a plant, and the group does not include *cyanobacteria* (a moneran that is casually—and misleadingly—called a "blue-green alga").

anaerobe   An organism adapted to low or zero levels of oxygen. Cf. *aerobe*.

anastomosis   Interconnection between veins in a leaf, blood vessels, or fungal hyphae. Also, the process of creating such interconnection.

angiosperm   A flowering plant such as an oak tree or tulip. Seeds develop enclosed within the fruiting structures on branches.

arborescent lycophyte   An extinct lycophyte of tree size; some forms reached over fifty meters in height.

arbuscle   Haustoria of a mycorrhizal fungus that form within a host plant root cell; arbuscles have a bush or tree shape.

Archaea   A proposed kingdom of bacteria, including methanogens, halobacteria, and some thermoacidophiles.

archaebacteria   See *Archaea*.

arthropleurid   An extinct relative of millipedes; grew to two meters length during the Paleozoic.

arthropod   Any member of the phylum Arthropoda, which comprises animals with jointed appendages and periodically molted exoskeletons. Living examples include lobsters, spiders, and insects.

asci   Plural of *ascus*.

ascomycetes   True fungi in which spores are formed in an *ascus*. Examples include mildew, truffles, baking yeast, and morels.

ascocarp   The fruiting body of an ascomycete fungus in which *asci* are formed.

ascospore   The sexually produced spore of an ascomycete fungus; formed in an *ascus*.

ascus (asci)   A saclike cell enclosing sexually produced spores. See *ascomycete*.

autotroph   An organism capable of biochemically manufacturing food from simple inorganic compounds. Cf. *heterotroph*.

bacterium (bacteria)   Any unicellular organism lacking nucleus and organelles. Bacteria are thus prokaryotic microbes belonging to the kingdom Monera.

basidiocarp   The fruiting body of a basidiomycete fungus; bears many *basidia*.

basidiomycetes   True fungi that produce spore-bearing basidia, some of which have the appearance of tiny upright tuning forks. Basidiomycete fungi include familiar forms such as the grocery store mushroom (*Agaricus bisporus*), *Coprinus* (the feces mushroom), *Amanita* (the "Angel of Death"), and the common shelf fungus *Polyporus* (figure 4.1).

basidiospore   A haploid spore of a basidiomycete fungus; usually borne in sets of four on the outside of a basidium. Spores form after a meiosis event.

basidium (basidia) A specialized reproductive hypha bearing, typically, four basidiospores on its tip. Basidia are found on the gills of a grocery store mushroom.

bdellovibrio A member of a group of parasitoidal bacteria. It invades a larger host bacterium, multiplies within, and eventually destroys the host bacterium, releasing numerous bdellovibrio progeny.

bifacial cambium A cylindrical layer of living cells within the stems of woody plants. New xylem cells are formed on the inner surface of the bifacial cambium, new phloem cells on its outer surface.

billion A thousand million.

biogenic Formed by life processes.

biogeophysical Pertaining to the interaction of living processes and geological forces on the surface of Earth. For instance, the uptake of carbon dioxide from the atmosphere by plants is a biogeophysical phenomenon.

biomass The total amount of living matter produced by organisms of a given region or organisms of a given type.

biotic Caused by life or otherwise associated with living organisms.

biotransfer The uptake and transport of chemical substances by the cells or body fluids of living organisms.

biramous Arthropod legs that have two leg branches emanating from a single joint socket. Cf. *polyramous, uniramous*.

blue-green alga Obsolete term for *cyanobacterium*.

bryophytes Nonvascular plants such as liverworts and mosses.

buffer Any substance which by its very presence within a medium tends to stabilize or preserve the physical characteristics (such as acidity and alkalinity) of the medium.

bulbil   A tiny plantlet formed in certain species of plants. Acts as an asexual propagule.

Calamites   An extinct genus of endarch, sphenophyte vascular plants (figure 6.28).

cambium   The layer, in vascular plants, of growing cells between the wood and the bark. See *bifacial cambium*.

Cambrian Explosion   The geologically brief interval (lasting perhaps only a few million years) at the beginning of the Paleozoic during which the majority of animal phyla appear in the fossil record. It is thought by many paleontologists to represent an interval of extraordinarily rapid evolution.

carbonaceous   Composed of carbon or of carbon-rich materials such as hydrocarbons. Carbonaceous fossils are generally a blackish color.

cation   An ion with a positive electrical charge (pronounced CAT-ion).

cellulose   A carbohydrate formed of tightly chemically bonded sugar molecules; the chief constituent of plant cell walls and fibers. Resistant to bacterial degradation.

cestode   A parasitic flatworm with no mouth or intestine.

charophyte   A multicellular, chlorophyte alga of the group that includes stoneworts and brittleworts.

chemoautotroph   Any autotrophic organism that uses *chemosynthesis* (as opposed to photosynthesis) to manufacture food.

chemosynthesis   The biochemical production of food from inorganic, energy-rich chemical compounds such as hydrogen sulfide or methane.

chitin   A rigid organic substance formed from sugars and protein; commonly utilized in the outer cuticle of animals.

chitinous   Formed of chitin.

chlamydospore   An asexual spherical structure of fungi formed by the swelling and partitioning of a living hyphal segment. The structure becomes isolated from its parent mycelium by a thick wall.

The fungus plasm hides in this structure during times of environmental stress, and emerges from it when beneficial conditions return.

chlorophyll   The main pigment required for photosynthesis; varieties are chlorophyll a, chlorophyll b, and chlorophyll c.

chlorophyte   An alga of the eukaryotic phylum Chlorophyta, the green algae. The phylum includes both unicellular (e.g., *Chlamydomonas*) and multicellular (e.g., charophytes, *Ulva*, *Halimeda*) examples. Starch is the primary storage material in cells. Probably ancestral to *vascular plants*.

chloroplast   An organelle capable of photosynthesis; found within cells of photosynthesizing eukaryotes such as plants. Originally derived from free-living *cyanobacteria*.

chromophytes   Unicellular or multicellular protoctists bearing *chlorophylls* a and c but lacking *chlorophyll* b. Unicellular forms are sometimes called golden yellow algae owing to the color of the plastids. Includes *chrysophytes* and *diatoms*.

chrysophyte   A member of phylum Chrysophyta, a group of chromophyte protoctists. Most forms have undulipodia with fine tinsel-like projections.

chytrid   A funguslike protoctist belonging to the phylum Chytridiomycetes. Many chytrids are microscopic, obligately intracellular parasites and thus are important constituents of Hypersea. Not true fungi because undulipodiated cells are formed as part of the life cycle.

cilium (cilia)   A short *undulipodium*.

class   A category in the taxonomic hierarchy intermediate between phylum and order.

cleistothecium (cleistothecia)   A flask-shaped, non-opening fruiting body in which *asci* and *ascospores* of an ascomycete fungus are formed.

Coal Age   The Carboniferous period.

coal ball   A limey (calcium carbonate) *nodule* formed in coal. Coal balls are prized for their tendency to preserve ancient plants and fungi in exquisite detail.

conidia   Fungal spores produced asexually (singular, *conidium*).

conifers   A large group of gymnosperm plants, mostly evergreens, which includes pine, redwood, fir, and spruce. Leaves are simple, often needlelike or scalelike.

convergent evolution   Evolution of similar body shape, bio-chemical capabilities, or other traits in unrelated types of organisms. The wings of birds, insects, and bats are similar in function and thus are examples of convergent evolution.

cordaites   An extinct, Paleozoic group of endarch trees with strap-shaped leaves (fig 6.36). Thought to be ancestral to modern coni-fers.

cryptoendolith   Any organism that lives just beneath the surface of porous rock. The rock pore habitat provides shelter from environ-mental extremes.

cryptogamic   Pertaining to seedless plants or plantlike organisms that reproduce by means of spores; term is also applied to soils whose surface is stabilized by *bryophytes, bacteria, algae,* and/or *lichens* in the absence of seed plants.

cryptomonad   A unicellular, photosynthetic protoctist belonging to the family Cryptophyceae. Most forms are able to move around with the help of undulipodia.

cryptophyte   See *cryptomonad.*

cutenized surface   Surface of a plant in which the outermost cells have been waterproofed by a coating of waxy and resinous substances.

cuticularized   See *cutenized surface.*

cyanobacterium   A photosynthetic bacterium distinguished by its blue-green color. See *blue-green alga.*

cycadeoids   An ancient, extinct group of cycadlike plants. Some forms had globular, squat stems.

cycadophytes   A plant group including cycads and cycadeoids.

cycads   A group of seed-forming plants with thick, stubby trunks and fernlike foliage.

cytoplasm   The living interior substance of a cell, exclusive of the nucleus.

definitive host   The final or main host of a parasitic organism having multiple hosts in its life cycle. Sexual reproduction of endoparasites occurs within the definitive host.

deuteromycetes   True fungi lacking any known method of sexual reproduction. This is an artificial grouping, consisting of asexual forms of both asomycete and basidiomycete fungi.

diatom   A photosynthetic unicellular protoctist; forms an opaline skeleton. Many forms are marine. Belongs to the *chromophyte algae*.

dictyostelous   Applied to plant stems composed of multiple *steles* bundled together to form a single stalk.

diffusion   As used in this book, the gradual intermingling of different types of molecules in a liquid.

dinoflagellate   A unicellular protoctist able to swim with the aid of undulipodia. Many forms are photosynthetic and many are also able to form a protective cyst.

diplobiontic   Having alternation of generations. The alternation is between a diploid and a haploid stage of the life cycle, with both the diploid and haploid stages developing into multicellular organisms. Cf. *haplobiontic*.

diploid   Having two copies of each chromosome. Most animal body cells are diploid, and the cells in a sporophyte plant are diploid. Cf. *haploid*.

disparity   A measure of taxonomic distance between distantly related organisms. For example, there is greater taxonomic disparity between a fish and a beetle than between a fish and a frog.

diversity   A measure of the number of different types of organisms in a particular place or at a particular point in time.

dolipore septum   A barrel-shaped structure, covered by a membranous pore cap, which perforates the septa between mycelial filament sections in a basidiomycete fungus.

dolomite A magnesium-rich sedimentary rock similar to limestone.

dorsal The back or upper side of an animal. Cf. *ventral*.

ecophysiology The study of the adaptations of organisms to their environments, particularly as regards their adaptations to new, challenging or stressful environments.

ecotype A variety or subpopulation within a species population that is adapted to particular ecological conditions.

Ediacaran biota A group of marine, megascopic fossils found in strata deposited before the Cambrian period. Thought to be fossil animals by some, others believe that these forms are unrelated to either plants or animals.

elater An elastic filament attached to the surface of a spore; facilitates dispersal by wind.

embryophyte An embryo-forming photosynthetic land organism, either vascular or nonvascular. Synonym: plant.

emergence See *enation*.

enation Leaflike, unorganized lump of tissue protruding from the stem of an early vascular plant. Presumed to function as a light-gathering organ.

endarch Any vascular plant in which the oldest, smallest-diameter xylem cells are at the center of the initial stele (figure 6.15). Most modern plants are endarch. Cf. *exarch*.

endemic Restricted to a particular region or locality.

endoparasite A parasite that lives within the body of its host.

endophyte A plant, fungus, or funguslike organism that is endosymbiotic or endoparasitic within a host plant.

endosperm The kernel of a seed; a "fraternal twin" of an angiosperm embryo, created during the same episode of fertilization that begins the embryo. The main function of the endosperm is to provision the embryo with a supply of nourishment.

endosymbiont   A symbiotic organism living within the tissues or the cytoplasm of another, larger organism called the host.

eocytes   According to some bacteriologists, the eocytes are members of a kingdom of bacteria including some but not all high-temperature, sulfur-metabolizing bacteria. (Others say eocytes are a phylum, not a kingdom.)

epidermis   The surface layer of a multicellular organism. May be strengthened by cuticle.

epiphyte   A parasitic plant or lichen that grows on the surface of a larger plant. Epiphytic plants may send roots into the tissues of the host plant and derive nutrients from the host.

eubacteria   Bacteria exclusive of *halobacteria, methanogens,* and some *thermoacidophiles.*

eukaryote   Any organism with a membrane-bound nucleus and organelles. Cf. *prokaryote.*

exarch   Any vascular plant in which the oldest and smallest-diameter xylem cells of the initial stele are on the outside margin, putting them closer to the stem epidermis (figure 6.15). Cf. *endarch.*

family   A category in the taxonomic hierarchy intermediate between class and genus.

fertilization   The joining of one male and one female gamete to form a diploid zygote.

flagellum (flagella)   A whiplike motility structure in a bacterium. Analogous but differently constructed structures in eukaryotes are called *undulipodia.*

fluke   See *trematode.*

foraminifer (foraminifera)   A particular kind of unicellular protoctist. Many forms are marine and secrete casings of calcium carbonate. They belong to phylum Granuloreticulosa.

fungus (fungi)   Any eukaryotic organism belonging to the kingdom Fungi. Such organisms are characterized by hyphae, formation of a mycelium, and lack of undulipodiated cells at any point in their life cycle. Protoctistan organisms with funguslike lifestyles are sometimes called fungi, although they are not true fungi.

gamete   Haploid reproductive cell of a eukaryote. For example, an unfertilized egg and a sperm cell are both gametes.

gametophyte   The haploid generation of a plant. The gametophyte gives rise to haploid gametes that fuse to form a diploid embryo. This embryo develops into the *sporophyte* generation.

genomic   Having to do with genes or their expression.

genus (genera)   A category in the taxonomic hierarchy intermediate between family and species. The first part (always capitalized) of a species' Latin name is its genus.

gigaton   One billion metric tons.

gnetales   A group of gymnosperm plants not closely related to conifers. Includes the xeric, shrubby Mormon Tea plant (*Ephedra*).

Gondwana   A supercontinent that formed before the Paleozoic and lasted into the Mesozoic era; it was composed of most of the present-day southern continents.

guard cells   Sausage-shaped cells occurring in pairs at the margins of plant stomates. These cells open the stomate when turgid and close the stomate when flaccid, thus helping the plant conserve water.

gymnosperms   Plants characterized by presence of cones. Gymnosperm seeds are considered "naked" because they lack endosperm and a protective seed coat. Examples include junipers, pines, and other familiar evergreens; also *cycads* and gingkos.

halobacteria   Bacteria tolerant of highly saline environmental conditions.

haplobiontic   Sexually reproducing organisms lacking alternation of generations (such as humans). The organism spends most of life in a diploid state; the only haploid cells in the life cycle are the sperm and ovum. No multicellular haploid generation develops. Cf. *diplobiontic*.

haploid   Having one copy of each chromosome. Cf. *diploid*.

haustorium (haustoria)   Any absorptive hyphal branching structure formed by parasitic or symbiotic fungi or funguslike organisms.
Haustoria penetrate host cells in order to exchange or absorb nutrients.

heterosporous   Applied to plants having two different kinds
(sizes) of spores, usually, megaspores that grow into female gametophytes and microspores that grow into male gametophytes (figure
6.30).

heterotroph   An organism dependent on other organisms as food
sources.

homologous   Evolutionarily derived from the same antecedent
structure. For example, the arm of a human and the wing of a bird
are homologous organs, sharing the same evolutionary derivation
from the forelimb of a reptile.

homosporous   Capable of forming only a single size of spore.

hydrogen sulfide H_2S, the rotten-egg gas.

hypermarine   Of, or pertaining to, *Hypersea*.

hyperparasite   A parasite of a parasite.

Hypersea   The sum total of eukaryotic life on land, plus any
symbionts (such as parasites) living within the tissues of eukaryotic
land organisms, be they other eukaryotes, bacteria, viruses or viruslike entities capable of living in and/or reproducing within the
tissues of a host. Hypersea includes cell walls, cell membranes,
cytoplasm, and cell-bearing or acellular body fluids of land eukaryotes.

hypha (hyphae)   A single filament of a fungus.

hypovirulence   Reduction in the severity of disease symptoms in a
host infected by a parasite, without apparent damage to the parasite.
Hypovirulence can be induced by hyperparasitism of the disease-
causing parasite.

ion   An atom or group of atoms bearing an electrical charge.

kingdom   The highest official category in the hierarchical classification of organisms.

leaf spot disease   A spot-forming infection of a leaf surface caused by several kinds of parasitic fungi or bacteria.

lichen   A composite organism formed by symbiosis of a photosynthetic *phycobiont* and a fungal or funguslike *mycobiont*.

lignin   A complex organic compound formed by plants; it adds stiffness and strength to the conducting cells in wood. Lignin is insoluble in water and is chemically quite unreactive.

lineage   An ancestor-descendant linkage through time; family tree.

littorinids   Air-breathing snails belonging to the families Littorinidae and Pomatiasidae.

lobopod   An arthropodlike animal with three or more pairs of unjointed legs on a cylindrical body. Claws occur at the leg tips.

lycophyte   An exarch plant, usually vascular, with sporangia closely associated with leaves or enations, in contrast to a *zosterophyll*, whose sporangia are not associated with its spines or other leaflike organs. Includes living clubmosses such as *Lycopodium* and *Selaginella* plus ancient arborescent lycophytes such as *Lepidodendron*.

lyginopterid   A type of *pteridosperm* (seed fern) vascular plant with an enclosed seed and a single stele in its stems. Cf. *medullosan*.

mass extinction   Extinction of many different types of unrelated organisms during a geologically brief interval of time.

medullosan   A type of *pteridosperm* (seed fern) with multiple steles in each stem. Cf. *lyginopterid*.

megafossil   Any fossil visible to the unaided eye.

megasporangium   A sporangium in which megaspores are formed.

megaspore   In a heterosporous plant, a large spore that can potentially develop into a (usually) female gametophyte. Cf. *microspore*.

meiosis   Reduction in chromosome number during cell division from the diploid number of chromosomes to the haploid number of chromosomes (i.e., half the diploid number). Most commonly occurs during the formation of gametes. Cf. *mitosis.*

metazoan   An animal with cells organized into tissues and organs.

methane CH_4, swamp gas.

methanogens   A group of bacteria that generate methane as a byproduct of metabolism.

microbial mat   A feltlike scum formed by aqueous bacteria. *Cyanobacteria* are common creators of microbial mats.

microbiota   A group of microbes living together in the same environment.

micron   One one-thousandth of a millimeter.

microsporan   A minute protoctist belonging to the phylum Microspora. Microsporans are all obligate intracellular parasites, 1–20 microns in size. Best known genus is *Nosema.*

microspore   In a heterosporous plant, a tiny spore that can potentially develop into a (usually) male gametophyte. Cf. *megaspore.*

milliOsmole   A unit used to measure osmotic pressure, abbreviated mOsm.

mitochondrion (mitochondria)   A food-metabolizing organelle in a eukaryotic cell. Thought to be derived from originally parasitoid, and subsequently endosymbiotic, bacteria.

mitosis   Cell division in which the number of chromosomes in daughter cells is the same as in the parent cell. Cf. *meiosis.*

Monera   The kingdom of prokaryotes; includes all bacteria.

monolete   Applied to spores with a single linear opening slit. Cf. *trilete.*

mutualism   A type of symbiosis in which both partners benefit from the association.

mycelium (mycelia)   The vegetative part of a fungus, formed by filaments called hyphae.

mycobiont   The fungal (or funguslike) member of a lichen symbiosis. Cf. *phycobiont*.

mycorrhiza   The symbiotic association of a fungus or a funguslike organism with the roots of a plant. The roots are invaded by or covered by hyphae; the hyphae play a role in absorption of nutrients from the soil.

myriapod   An arthropod with uniramous, jointed appendages. Centipedes and millipedes are myriapods.

nematoclast   A type of fossil that is microscopic, tubular, and composed of organic matter. Thought to be derived from early land organisms (such as nematophytes) or from ancient aquatic algae.

nematode   A wormlike, minute (often less than one millimeter in length) animal abundant in soil and other environments. Nematodes are commonly parasitic.

nematophyte   An enigmatic Paleozoic organism that lived on land or in shallow water and that was composed of hypha-like filaments and/or tubes. *Prototaxites* is the best-known Devonian example of a nematophyte.

nodule   A lump or spheroidal mass of mineral crystals, contrasting in composition with the surrounding rock in which it formed.

nucleus (nuclei)   The membrane-bound structure in the cytoplasm of a eukaryotic cell; contains the chromosomes.

nutricline   The oceanic depth above which too few nutrients are available for the growth of many types of photosynthetic organisms. Nutrients are depleted above this horizon because of the amount of photosynthetic activity in well-lit ocean surface waters.

onychophoran   An extant lobopod; also called a velvet worm.

oogonium (oogonia)   A unicellular structure containing ova.

oomycete   A funguslike protoctist that differs from true fungi by having an undulipodiated life stage.

order   A category in the taxonomic hierarchy narrower than class and broader than family.

organelle   A discrete membrane-bound structure within the cyto-plasm of a eukaryotic cell, such as a *chloroplast*.

osmoconformer   An organism that does not actively regulate the inorganic solute composition (osmotic pressure) of its blood.

osmoregulator   An organism that actively regulates the inorganic solute composition (osmotic pressure) of its blood.

osmosis   The tendency for water to diffuse through a semiperme-able membrane (i.e., through a membrane permeable to water but not to solutes dissolved in water) into a solution of higher concentra-tion, thus equalizing the concentrations of solutes on both sides of the membrane.

osmotic pressure   The force exerted by water passing through a semipermeable membrane by *osmosis*.

ovum (ova)   The unfertilized female gamete.

Oxygen Revolution   That point in geologic time, approximately two billion years ago, when significant quantities of free oxygen began to accumulate in the atmosphere and the sea. Also called the Oxygen Crisis.

paedomorphic   Retaining juvenile characteristics into adulthood or failing to develop typical adult features after the onset of sexual maturity.

palynoflora   A collection of palynomorphs that lived at the same time or in the same environment.

palynology   The study of microfossils preserved as organic com-pounds. Usually such fossils (*palynomorphs*) are freed from the rock matrix by dissolving the rock in acid.

palynomorphs   Organic-walled microfossils.

parasitism   A form of *symbiosis* in which one organism lives at the expense of the other, contributing nothing beneficial to its host.

parasitoid   A parasite that completes its entire life cycle within a single host, usually resulting in the destruction of the host.

period   A major interval of geologic time; see figure 1.1.

permineralization   The addition, during fossilization, of mineral deposits to the organic remains being fossilized.

pH   A symbol used, in combination with a number between 1 and 14, to denote the alkalinity or acidity of a solution. Low pH values are acidic and high values are alkaline.

phloem   Plant tissue consisting of conductive cells that carry food to the nonphotosynthesizing parts of a vascular plant.

photoautotroph   An autotroph that creats its substances from simple inorganic compounds with the aid of solar energy.

phycobiont   The photosynthesizing algal or cyanobacterial member of a lichen symbiosis. Cf. *mycobiont*.

phylum (phyla)   A category in the taxonomic hierarchy intermediate between kingdom and class.

pith   A plant tissue formed of thin-walled living cells in the center of mature steles. Air spaces may be present between cells.

placer   A stream or river deposit of concentrated heavy minerals such as gold.

plasm   The liquid part of a cell interior or organism's body.

plasmodiophoromycetes   A group of parasitic funguslike organisms belonging to phylum Plasmodiophoromycota. Most are endoparasites.

plastid   An enclosed structure or organelle in the cytoplasm of eukaryotic cells that contains oil, starch, or pigment.

pollen   A haploid male gametophyte enclosed within a resistant coat. Pollen grains serve as the male sex cells of flowering plants.

polyramous   Arthropod appendages with multiple branches; Cf. *uniramous, biramous*.

Porphyridiales   An order of red algae, a type of photosynthetic multicellular protoctist.

predation pressure   Any difficulties imposed on prey organisms by their predators, particularly those that lead to evolutionary changes in the prey.

progymnosperm   An extinct group of early tree-forming, woody endarch plants thought to be related to the ancestors of modern gymnosperms. Examples include the Devonian *Aneurophyton*, *Archaeopteris*, and *Eospermatopteris*.

prokaryote   Any unicellular organism without a membrane-bound nucleus and without organelles. Synonyms: *bacterium*, *moneran*.

propagule   Any sexually or asexually produced reproductive structure such as a seed, spore, or egg that disperses outside the body of its parent to create a new individual.

Proterozoic   An approximately two-billion-year era of geologic time that immediately preceded the Paleozoic era.

protist   A unicellular eukaryote.

protoctist   Any eukaryotic organism that is neither plant, animal, nor fungus. May be unicellular or multicellular.

pteridosperms   An extinct group of endarch plants that first appeared in the early Devonian and died out in the Jurassic. Also called seed ferns.

radiolarian   A kind of predaceous, unicellular marine protoctist.

refugium (refugia)   Any bounded environment enabling a species of organism to avoid total extinction. Organisms that inhabit a refugium once had wider geographic distribution.

regression   (1) In geology: withdrawal of seawater from the land surface; a lowering in sea level. (2) In statistics: the "best-fit" line calculated from an array of data.

rhizoid   Any filamentous rootlike structure lacking vascular tissue. May be multicellular or unicellular.

rhizome   A multicellular, leafless horizontal stem; often forms underground.

rhodophyte   A member of a protoctist phylum; also called red algae. Most forms are marine, and undulipodia are absent from all life stages.

rhyniophyte   A member of a group of Silurian and Devonian vascular(?) plants. Rhyniophytes are among the earliest and simplest known plants, consisting of slender leafless stems with terminal (stem-tip) sporangia. Best known genus is *Rhynia*.

rock varnish   A darkly colored coating of manganese and iron oxides formed on rock surfaces by the activity of bacteria.

Rodinia   A Proterozoic supercontinent consisting of all or nearly all of the continents; this supercontinent broke into continents by plate tectonics and continental drift about 700–500 million years ago.

roundworm   See *nematode*.

saprotroph   An organism that utilizes the already dead remains other organisms for food.

saturation vapor pressure   The amount of water vapor that air is able to hold at a particular temperature. As air temperature falls, saturation vapor pressure drops and dew forms.

sclerite   An isolated piece of a multielement skeleton that once covered and protected the body surface of an animal that lived in the water.

seed   A fertilized megasporangium wrapped in a covering, or integument.

seed ferns   See *pteridosperms*.

septum (septa)   A dividing wall or membrane.

septate   Partitioned into compartments by septa.

silicate   A rock or mineral consisting primarily of silicon and oxygen.

soredium (soredia)   A lichen propagule consisting of an algal cell wrapped by fungal filaments.

species   A group of organisms capable of interbreeding (or, from fossil evidence, inferred to have been able to interbreed); this is the "lowest" or fundamental unit of classification in the taxonomic hierarchy, ranking just below *genus*.

sphenophyte   An endarch plants characterized by ribbed, jointed stems and whorled branches. Modern horsetails are a living example.

spirochete   A bacterium with a corkscrew shape.

sporangium (sporangia)   A sac in which spores are formed.

spore   Any propagule with a resistant coating capable of developing into a new adult organism.

sporophyte   The diploid generation of a plant or protoctist; gives rise to haploid spores that develop into the *gametophyte* generation.

stele   A central cylinder or vascular bundle of *xylem* and *phloem* in the stems and roots of plants.

stipe   A relatively thin stalk.

stomate   A tiny intercellular pore in plant epidermis. *Guard cells* flank the opening to the outside, and regulate the amount of water leaving the plant.

strata   Layered sedimentary rocks.

stratigraphy   The study of the age, correlations, and depositional environments of ancient strata.

symbiogenesis   The evolutionary creation of a new life form by the permanent fusion, via symbiosis, of two or more, once free-living, distantly related species of organisms.

symbiosis (symbioses)   An intimate association between organisms belonging to different species.

taxonomy   The science of categorizing organisms into groups based on shared anatomical structures or other features indicative of common descent.

telome   A theoretical unit of branch organization consisting of the most distal (outward) twig of a many-forked (dichotomously branched) branch.

thallus (thalli)   The body of a plant or photosynthetic plantlike organism lacking roots, leaves, or stem. Often flattened; may branch into lobelike structures.

thermoacidophile   A bacterium tolerant of hot, acidic conditions.

thermophile   A bacterium tolerant of hot environmental conditions.

thin section   A slice of a rock or a fossil mounted to a glass slide with epoxy and ground and polished so thin (usually 30–50 microns) that light can pass through, enabling study by microscope.

tracheid   A water-filled, tubular cell, tapered at both ends and strengthened by bands of lignin. Tracheids are the conductive cells of xylem in vascular plants.

transgression   Landward motion of the shoreline; rise in sea level.

trematode   A type of parasitic flatworm; fluke.

trichomycete   A symbiotic fungus found only within the gut of myriapods and in certain other arthropods.

trilete scar   A triradial marking on the surface of a spore that was once part of a spore tetrad or cluster of four spores in a tetrahedral arrangement. The three arms of the scar indicate the edges where the spore was attached to the other three spores of the tetrad.

trimerophyte   A Devonian plant that bore lateral branches but no leaves. Thought to have been ancestral to ferns and sphenophytes. Example: *Psilophyton* (figure 6.10).

tritiated   Containing tritium, a radioactive isotope of hydrogen.

trophic   Having to do with food or feeding.

ultrastructure   Structure so fine as to be visible only with the aid of electron microscopy.

undulipodium (undulipodia)   The whiplike motility organelle of a eukaryotic cell; has a distinctive internal structure. Not to be confused with the bacterial *flagellum,* which is not an organelle and has an entirely different structure.

uniramous Arthropod legs having one leg branch emanating from a single joint socket. Cf. *polyramous, biramous.*

upwelling, hypermarine Delivery of nutrients from soil to photosynthetic living tissue by vascular plants and their symbionts.

upwelling, marine Delivery of nutrients from deep ocean waters to ocean surface waters by upwardly directed currents driven by physical processes such as wind.

vacuole A droplet of fluid within a cell.

vascular plant A plant with tissues such as *xylem* and *phloem* capable of conducting water and dissolved nutrients from one part of the plant to another.

vascular tissue The fluid-transporting tissue of a vascular plant.

vegetative Actively growing into a multicellular organism; nonreproductive.

Vendian The late Proterozoic geologic period immediately preceding the Cambrian period.

ventral The "belly" or underside of an animal. Cf. *dorsal.*

Volvocales An order of eukaryotic green algae.

xeric Adapted to arid conditions on land.

xylem The water conductive tissue of vascular plants; xylem carries water and mineral salts from the soil to the upper reaches of the plant. The woody tissues of plants are composed of layers of *xylem tracheids.*

zoosporangium (zoosporangia) A sporangium bearing zoospores.

zoospore A unicellular, swimming propagule formed by oomycetes, chytrids, some plasmodiophoromycetes, and other protoctists. Spermlike in appearance but unable to undergo sexual fusion. Instead, they are able to transform into a subsequent, often larger, developmental stage of the organism.

zosterophyll A member of an extinct Devonian group of exarch plants (figure 6.14). Some forms have kidney-shaped sporangia placed symmetrically along the sides of the stem. May have been ancestral to the *lycophytes*.

zygomycetes   True fungi in which most of the mycelia lack septa. Sexual reproduction is accomplished by fusion of two hyphal filaments to form a zygote.

zygote   A cell or embryo formed by the fusion of two gametes or sex cells.

FIGURE SOURCES AND CREDITS

MHC refers to Mount Holyoke College; MCZ refers to Museum of Comparative Zoology, Harvard University; numbers are the sample numbers of the specimens stored at these institutions. Unless otherwise attributed, photographs and drawings are by the authors.

Figures 1.2 and 1.3. Photo courtesy of T. N. Taylor, W. Remy, and H. Hass. Same specimens are illustrated in T. N. Taylor, W. Remy, and H. Hass, 1992, Fungi from the Lower Devonian Rhynie Chert: Chytridiomycetes, *American Journal of Botany* 79:1233–1241.

Figure 1.4. Specimen discovered by Heather Winkelmann. Sample 8c of 6/15/93.

Figure 1.5. Photo courtesy of Julia Casperd, Gay C. Marris, and Douglas Speirs. Same specimen illustrated in D. C. Speirs, T. N. Sherratt, and S. F. Hubbard, 1991, Parasitoid diets: Does superparasitism pay? *Trends in Ecology and Evolution* 6:22–25.

Figures 1.6 and 1.7. Scanning electron photomicrographs by P. Weaver and M. Rice.

Figure 1.8. From M. A. S. McMenamin and D. L. S. McMenamin, 1993, Hypersea and the Land Ecosystem, *BioSystems* 31:145–153.

Figure 3.2. Redrawn by H. Winkelmann from figure 7A of J. G. Gehling, 1988, A cnidarian of actinian-grade from the Ediacaran Pound Subgroup, South Australia, *Alcheringa* 12:299–314.

Figure 4.1. Photograph taken by H. Winkelmann on the Mount Holyoke College campus.

Figure 4.2.   Photograph taken at 28 Westbrook Road, South Hadley, Massachusetts on June 11, 1993.

Figures 4.3–4.5.   Photographs courtesy of John R. Watterson.

Figures 4.6 and 4.7.   Photographs courtesy of M. Rice and G. Klungness.

Figure 4.8.   Sketch by H. Winkelmann.

Figure 4.9.   Sketch based on N. A. Volkova, 1976, O nakhodke Dokembriishikh spor s tetradnim rubtsom, *Paleontologiya Morskaya Geologiya Mezhdunarodny Geologicheskiya Kongressa, XXV Session,* Akademiya Nauk SSSR, Ministerstvo Geologichestvo SSSR, "Nauka," pp. 14–18.

Figure 5.1.   Figure courtesy of S. Ostoumov. See also p. 31 of I. I. Mochalov, ed., 1988, *Prometei, Istoriko-biograficheskii al'manakh serii Zhizn' zamechatel'nykh lyudei, 125-letiyu so dnya rozhdeniya V. I. Vernadskogo posvyashchaetsya,* Moscow: Molodaya gvardiya.

Figure 6.2–6.8.   Photos courtesy of P. K. Strother.

Figure 6.9.   Sample R1(E) of 3/31/92; Rhynie Chert.

Figure 6.10.   *Psilophyton* cf. *P. princeps* Dawson, Devonian (Portage-Ithaca shale), Ithaca, New York, MHC 1645.

Figure 6.14.   Adapted from P. Kenrick and D. Edwards, 1988, A new zosterophyll from a recently discovered exposure of the Lower Devonian Senni Beds in Dyfed, Wales, *Botanical Journal of the Linnean Society* 98:97–115.

Figure 6.19.   *Leclercqia,* MHC 1639.

Figure 6.21.   *Lepidodendron aculeatum* Sternberg, MHC 8823.

Figure 6.23.   *Lepidophloios,* sample 1 of 3/1/79; latex peel from an external mold in the collection at Stanford University.

Figure 6.24.   *Sigillaria ichthyolepsis* Sternberg, MHC 2375.

Figure 6.25.   Figure modified from H. P. Banks, 1972, *Evolution and Plants of the Past,* London: MacMillan.

Figure 6.27. *Annularia radiata* (Brongniart) Sternberg, Carbondale Formation, Mazon Creek, Illinois, MHC 3222.

Figure 6.28. Sample number is MHC 1424.

Figure 6.29. *Sphenophyllum* acetate peel from a Carboniferous coal ball (unknown locality), sample 28 of 6/15/93. Probably prepared by F. D. Reed.

Figure 6.30. Figure modified from H. P. Banks, 1972, *Evolution and Plants of the Past*, London: MacMillan.

Figure 6.31 and 6.32. Previously unpublished figures courtesy of W. G. Chaloner.

Figure 6.34. *Stigmaria ficoides* var. *Goepp*, MHC 2784.

Figure 6.35. American Lotus, *Nelumbo lutea*, collected live at Forest Park, Springfield, Massachusetts in 1992.

Figure 6.39. *Neuropteris*, McMenamin collection K2P13N3. Locality unknown.

Figure 6.40. *Neuropteris decipiens*, Carbondale Formation, Grundy County, Illinois, MHC 1422.

Figure 6.41. *Neuropteris loschii* Brongniart, MHC 1999.

Figure 6.42. Placement of floral provinces adapted from W. G. Chaloner and S. V. Meyen, 1973, Carboniferous and Permian floras of the northern continents, in A. Hallam, ed., *Atlas of Palaeobiogeography*, pp. 169–186. Amsterdam: Elsevier.

Figure 6.43. *Coleochaete*, same specimen as shown in figure 3 of L. E. Graham and L. W. Wilcox, 1983, The occurrence and phylogenetic significance of putative placental transfer cells in the green alga *Coleochaete, American Journal of Botany* 70:113–120.

Figures 6.44–6.46. Courtesy of J. G. Duckett and R. Ligrone. From J. G. Duckett and R. Ligrone, 1992, A light and electron microscope study of the fungal endophytes in the sporophyte and gametophyte of *Lycopodium cernuum* with observations on the gametophyte-sporophyte junction, *Canadian Journal of Botany* 70:58–72.

Spongiophyton reconstruction of figure 6.44 is based on specimens described in P. G. Gensel, W. G. Chaloner, and W. H. Forbes, 1991, *Spongiophyton* from the late Lower Devonian of New Brunswick and Quebec, Canada, *Palaeontology* 34:149–168.

Figures 6.49–6.52.   Photographs courtesy of P. Gensel.

Figure 6.53.   Drawn by H. Winkelmann.

Figures 6.54 and 6.55.   Photographs courtesy of P. K. Strother.

Figure 6.56.   Specimen from collection of S. Rachootin.

Figure 6.57.   *Lepidocaris*, Sample R1 of 7/12/93.

Figure 6.58.   Photograph by H. Winkelmann.

Figure 6.62.   Adapted from a photograph courtesy of Donald M. Ball.

Figure 7.2.   From M. A. S. McMenamin and D. L. S. McMenamin, 1990, *The Emergence of Animals: The Cambrian Breakthrough*, New York: Columbia University Press.

Figure 7.3.   From the Puerto Blanco Formation of northwestern Sonora, Mexico; Lower Cambrian. Field sample 5.5 of 12/17/82.

Figure 7.4.   *Euphoberia armigera?* Meek and Worthen, Type specimen, MCZ 5454/2b.

Figure 7.5.   *Acantherpestes major* Meek and Worthen, MCZ 7437/2.

Figure 7.6.   Scolopendromorph centipede from Takaragaike Park, Kyoto, Japan, sample 1 of 8/27/92.

Figure 7.7.   Sample 6 of 11/2/92.

Figure 7.9 and 7.10.   *Argulus* w.m. Ward's Natural Science Establishment 92 W 6020.

Figure 7.11.   *Cyclus americanus* from Mazon Creek, Illinois. Counterpart (inverse relief). Mount Holyoke College sample number pending.

Figure 8.2 and 8.3.   Chalcid wasps, sample 6 of 11/2/92.

Figure 8.4. Specimen is from probable *Hymenaea* amber from Cenozoic Dominican amber. The leaf (probably genus *Hymenaea*) was infected with fungal leaf-spot disease. Dominican amber is at least early Miocene in age (approximately 23 million years old). Specimen is from an amber necklace.

Figure 9.1. *Septobasidium* sketch inspired by the frontispiece of J. N. Couch, 1938, *The Genus Septobasidium*, Chapel Hill: University of North Carolina Press.

INDEX

Numbers in italics refer to pages with illustrations.

Designer: Linda Secondari
Text: 11/14 Caslon Old Style
Compositor: Maple-Vail
Printer: Maple-Vail
Binder: Maple-Vail